A History of Chemistry

INTERNATIONAL CHEMICAL SERIES

H. P. TALBOT, PH. D., CONSULTING EDITOR

A HISTORY
OF
CHEMISTRY

(Frontispiece)

ROBERT BOYLE
1627–1691

A HISTORY
OF
CHEMISTRY

BY

F. J. MOORE, Ph. D.

PROFESSOR OF ORGANIC CHEMISTRY IN THE MASSACHUSETTS INSTITUTE OF TECHNOLOGY.

First Edition

McGRAW-HILL BOOK COMPANY, Inc.
239 WEST 39TH STREET. NEW YORK

LONDON: HILL PUBLISHING CO., Ltd.
6 & 8 BOUVERIE ST., E.C.
1918

THE MAPLE PRESS YORK PA

PREFACE

This volume is the outgrowth of a series of talks which the author has for several years given to his students at the Institute of Technology, the hearers being members of the senior class specializing in chemistry, and hence familiar with its more important facts and principles. The lectures have dealt in a direct informal way with the fundamental ideas of the science: their origin, their philosophical basis, the critical periods in their development, and the personalities of the great men whose efforts have contributed to that development.

Put in book form the material has inevitably been somewhat expanded, and just as inevitably its presentation has assumed a more formal tone, without, it is hoped, losing all its spontaneity. Here, as before, the person addressed is the more mature student of chemistry, though it is believed that few portions of the book will present serious difficulties to the general reader. The aim has been to emphasize only those facts and influences which have contributed to make the science what it is today; hence such topics as the chemical achievements of the ancients and the history of alchemy have been compressed beyond the point which the tastes and inclinations of the writer might alone have dictated. In the discussion of later work, also, the claim of a topic for consideration has been not its practical but its historical importance. It has been asked, not whether the work was itself of value, but did it contribute a new fundamental idea. For this reason, to cite a single instance, the work of Werner on the metal-ammonias has been discussed at some length, while that of Emil Fischer on the sugars has been dismissed with a single word. Some modern topics, also, like the work of Werner just mentioned, or that of Bragg upon X-ray spectra have been treated in considerable detail because they lie outside the field familiar to most undergraduates.

Little attention has been paid to questions of priority. A great discovery is usually preceded by a multitude of earlier observations, the sum total of which may even include all the fundamental facts involved. Hence arise the familiar troubles encountered by the conscientious student when he attempts to learn who invented the steam engine or who really discovered America. We can, however, save ourselves most of these difficulties if we reflect that from the historical standpoint the discoverer of a great truth is usually the one through whose efforts it first becomes available to the race.

A word may not be out of place concerning the illustrations. These have been reproduced from many sources, and have been selected entirely for their historical interest without regard to their artistic merit. It will therefore occasion no surprise that they differ greatly among themselves in the latter respect. Thanks are especially due to Professor Derr of the Institute for placing at the disposal of the author his unusual skill and knowledge of photography, and also to several publishers who have kindly permitted the use of copyrighted material.

The value of the historical method for studying every department of human thought is now so universally recognized that it requires no emphasis, but to the younger student of chemistry it may not be superfluous to point out that, by observing the errors and misunderstandings of the past, we learn to avoid errors in our own thinking; by acquaintance with the way in which great men have solved problems, we are assisted in solving problems of our own; by observing the different aspects presented by the same facts in the light of successive theories, we acquire an insight obtainable in no other way into the nature, limitations and proper function of all theories. Finally, as we study how man's knowledge of nature has broadened and deepened with the years, we acquire a better understanding of the trend of thought in our own times, and of the exact bearing of each new discovery upon the old but ever recurring problems of the science. At no period has the development of chemistry been more rapid or more interesting than it is today, and the author indulges the hope that even this brief sketch of its history may assist the reader to follow that development with a fuller appreciation of its significance, for, after all, we study the past that we may understand the present and judge wisely of the future.

F. J. MOORE.

MASSACHUSETTS INSTITUTE OF TECHNOLOGY.
April, 1918.

CONTENTS

PAGE

PREFACE v

LIST OF PLATES xiii

CHAPTER I

CHEMISTRY AMONG THE ANCIENTS 1
Ancient Arts; Theories of the Philosophers; Thales; Anaximenes and Leucippus; Heraclitus and Empedocles; Democritus and the Atomic Theory; Aristotle and the Four Elements; Archimedes and Eratosthenes; Pliny.

CHAPTER II

CHEMISTRY IN THE MIDDLE AGES. ALCHEMY 9
Intellectual Decline During the Middle Ages; Rise of Alchemy; Alchemistic Traditions; Fundamental Ideas; Practical Achievements; Prominent Individuals; Decay of Alchemy.

CHAPTER III

CHEMISTRY IN THE RENAISSANCE 18
The Revival of Learning; Career of Paracelsus; His Fundamental Ideas; Agricola; Van Helmont; Glauber.

CHAPTER IV

BOYLE AND HIS CONTEMPORARIES. THE PHLOGISTON THEORY 25
Boyle; Mayow and Hales; Kunkel and Becher; Stahl and the Phlogiston Theory; Advantages and Faults of the Theory; Early Criticism; Hoffmann, Boerhave, Marggraf; Geoffroy and His Tables of Affinity; Rouelle.

CHAPTER V

THE LATER PHLOGISTIANS. THE DISCOVERY OF OXYGEN 33
Black's Work on Magnesia Alba; Cavendish; Life and Work of Scheele; His more Prominent Investigations; Book on Air and Fire; Priestley; His Discovery of Oxygen.

CHAPTER VI

LAVOISIER 47
Life and Character; Early Studies; Gain in Weight by Combustion;

PAGE

Conception of Chemical Composition; The Dualistic System; Berzelius's Interpretation of Electrolysis; Wöhler's Reminiscences of his Studies with Berzelius.

CHAPTER XII

UALISM IN ORGANIC CHEMISTRY 120
Wöhler, Liebig, and Dumas; Wöhler; Life and Character of Liebig; The Laboratory at Giessen as a Center of Chemical Inspiration; Organic Analysis; Friendship of Liebig and Wöhler; Dumas; State of Organic Chemistry in 1828; The Vital Force and the Synthesis of Urea; The First Radical Theory; Ethylene; The Benzoyl Radical; The Ethyl and Acetyl Theories; Increasing Indefiniteness of the Radicals.

CHAPTER XIII

'HE REACTION AGAINST BERZELIUS 136
Substitution in Organic Chemistry; The Preparation of Trichloroacetic Acid; Dumas and the Idea of Types; Hostility of Berzelius; His Theory of Conjugate Compounds; Graham and Liebig on the Polybasic Acids; Revival of the Hydrogen Theory; Dumas's Work on Vapor Densities; His Inability to Accept Avogadro's Hypothesis; Polymorphism and Isomorphism; Irregularity of the Specific Heats; Faraday's Law; The Foundation of Electro-chemistry; Its Bearing upon the Origin of the Current; Criticism of Berzelius.

CHAPTER XIV

ERHARDT AND THE CHEMICAL REFORMATION. WILLIAMSON. 150
Laurent; Life of Gerhardt; Chemical Notation in 1850; Gerhardt's Atomic Weights; Distinction between Atoms and Equivalents; The Second Radical Theory; Criteria for the Basicity of Acids; Gerhardt's Text Book; Work of Williamson on Ethers; The Second Type Theory; Life of Williamson; Wurtz and Hofmann.

CHAPTER XV

'HE TRANSITION FROM THE TYPE THEORY TO THE VALENCE THEORY . . 164
Kolbe; Frankland; The Development of the Conjugate Compounds of Berzelius; The Metal Alkyls; Frankland's Anticipation of the Valence Theory; Kekulé; Multiple and Mixed Types; The Methane Type; The Quadrivalence of Carbon; Direct Connection between Carbon Atoms; Couper's First Graphic Formulæ; Kekulé's Text Book; The Vindication of Avogadro's Hypothesis; Anomalies Explained by Dissociation.

CHAPTER XVI

PA

THE PERIODIC LAW . . . . 1'

Revision of the Atomic Weights by Stas; Döbereiner's Triads; Arrangements of the Elements by Pettenkofer and Dumas; Systems of Gladstone, Cooke, and de Chancourtois; Newland's Law of Octaves; Lives of Lothar Meyer and Mendelejeff; System of the Latter; His Predictions; Anomalies of the Theory.

CHAPTER XVII

BUNSEN, BERTHELOT AND PASTEUR . . . . 1!

Life of Bunsen; The Work on Cacodyl; Gas Analysis; Geological Studies in Iceland; Theory of Geyser Action; Iodimetry; The Photochemical Investigations; The Spectroscope; Bunsen as a Teacher; Berthelot; Studies in Organic Syntheses; The Velocity of Esterification; Acetylene; The Law of Maximum Work; Explosives; Historical Studies; Religious Views; Pasteur; Studies in Molecular Asymmetry; Fermentation; Spontaneous Generation; The Disease of Silk Worms; Anthrax; Hydrophobia; The Pasteur Institute.

CHAPTER XVIII

ORGANIC CHEMISTRY SINCE 1860 . . . . 2]

Kekulé's Theory of the Benzene Ring; Van't Hoff; Stereoisomerism; Bivalent Carbon; Trivalent Carbon; Tautomerism; Special Researches; The Coal Tar Industry.

CHAPTER XIX

INORGANIC CHEMISTRY SINCE 1860 . . . . 2:

Stagnation during the First Years of this Period; Work of Moissan on Fluorine; The Electric Furnace; V. Meyer on Vapor Density at High Temperatures; The Dissociation of Iodine; Werner on the Metal-Ammonias; Stereoisomerism and Optical Activity among Inorganic Compounds; Work of Rayleigh and Ramsay on the Rare Gases of the Atmosphere; The Place of the New Elements in the Periodic System.

CHAPTER XX

THE RISE OF PHYSICAL CHEMISTRY 23

Early Physical Chemists; The Law of Mass Action; Wilhelmy; Berthelot; Guldberg and Waage; Willard Gibbs and the Phase Rule; Early Views Concerning the Mechanism of Electrolysis; The Ions of Faraday; Work of Hittorf and Kohlrausch; Studies of Raoult on the Freezing Point of Solutions; Osmotic Pressure; Its Interpreta-

PAGE

tion by Van't Hoff; Ostwald on the Affinity Constant of Organic Acids; Statement of the Law of Electrolytic Dissociation by Arrhenius; Influence of Physical Chemistry upon the Science as a Whole; Life and Activity of Ostwald; His Vindication of the Chemical Theory of the Galvanic Current.

CHAPTER XXI

ADIOACTIVITY. ITS INFLUENCE UPON THE ATOMIC THEORY 252

The X-rays; Experiments by Becquerel; Discovery of Polonium and Radium by Madame Curie; Rutherford's Work on Thorium; His Theory of Atomic Disintegration; Character of the Products; Helium from Radium; Details of Radioactive Transformation; Light Thrown by Radioactivity upon the Nature and Dimensions of the Atom; Genetic Relationships of the Radioactive Elements; Isotopy; X-ray Spectra; Their Use for Studying the Arrangement of Atoms in Crystals; Moseley's Comparison of Different Spectra; The Atomic Numbers and the Periodic Law.

IDEX 273

LIST OF PLATES

A HISTORY
OF
CHEMISTRY

CHAPTER I

CHEMISTRY AMONG THE ANCIENTS

A practical knowledge of many important chemical operations must have preceded the dawn of connected history. The preparation of wine and vinegar, the arts of pottery, elementary metallurgy, glass-making and dyeing are referred to as familiar processes in the earliest human records. Those who practised these arts in ancient times, however, recognized no bond of union between the various pursuits, and would themselves have been astonished if they had been classified together. The dyer, the potter, or the worker in metals either inherited his craft or else acquired skill through years of apprenticeship to some successful master. In either case the practical rules of procedure must have been handed down in the form of oral tradition since the artisan class was practically illiterate. From the number and variety of the industries successfully carried on, the sum total of knowledge of chemical phenomena involved in them must have been considerable, but we have no records of the details, and what we know of these early conditions depends upon the chance allusions of contemporary writers or upon the products unearthed by archaeological research.

Inscriptions representing glass-makers at work are found in Egyptian monuments of the XIth dynasty showing that considerable skill in this art must have been attained at least two thousand years before Christ. Indigo seems to have been used in dyeing ten centuries before the Christian era, while it is well known that iron and copper implements are found in remains entirely prehistoric.

References to "vinegar," "nitre" and "fullers sope" in the Bible show that these materials were in use at the time the books concerned were written, although we cannot always be sure that the names thus rendered in the translation signified just these substances in the original. Finally the scriptural account of certain experiences of Noah makes it clear that the preparation of intoxicating beverages at least preceded the writing of Genesis.

Fragmentary details of this sort might be multiplied indefinitely but this would not lead us to any helpful generalizations, for we have no records from which we might learn what the men who did the work thought about the processes involved. Probably they had no theories, since the field of each worker was too narrow for this.

The Greek Philosophers.—If the practical artisan left no theories, the same cannot be said of the speculative philosophers; in fact, we find some of the greatest minds of antiquity busying themselves with the fundamental nature of matter and discussing some of the material transformations which we should now classify as chemical.

It may be said at the outset that we can expect nothing from these men which might even serve as a foundation for a chemical philosophy. They had no first-hand knowledge of chemical transformations. Their social position kept them out of touch with those who might have given them practical information, and the whole atmosphere of the age discredited experiment as it discredited manual labor. Pure thought was alone held worthy of the philosopher, and by its means the Greeks made wonderful advances in mathematics and metaphysics. It was not surprising, therefore, that with no other guide they should have approached the problems of natural science with ill-founded confidence. Here, however, the costly experience of later centuries was destined to show that thinking which is not constantly checked by experiment leads only to unreliable results.

In spite of all this the great Greek philosophers exercised unbounded influence upon their contemporaries, and their ideas held sway throughout the Middle Ages. Indeed it is not difficult to detect some Greek influence upon scientific thinking even in the nineteenth century. For this reason their opinions must be considered at least in outline here.

Thales.—The first of the Greek philosophers of whom we have any record is Thales of Miletus who is supposed to have lived between 640 and 546 B.C.[1] He left no writings which have come down to us and we are indebted to Aristotle for most of what we know concerning him, as well as concerning several other of the earliest thinkers.

Thales seems to have established some important theorems in geometry and to have made certain astronomical observations; thus he is said to have determined the number of days in the year, and to have estimated the sun's angular diameter at $\frac{1}{720}$ of the zodiac. He is also credited with having predicted the date of an eclipse. He is best remembered among chemists for his statement that water is the origin of all things, but we can hardly be justified in receiving this statement in a literal sense. We have to recall that philosophers have always delighted in pointing out the ephemeral character of all things temporal, and this general condition of flux and change is well symbolized by water. If we add that Thales, as the inhabitant of a small island, was doubtless well acquainted with the sea and had some realization of its teeming life, he may well have grasped the idea that all life must have had its rise there.

Anaximines and Leucippus.—Anaximines and Leucippus were also citizens of Miletus who lived a little later than Thales, and are said to have held that air and earth respectively were the fundamental elements, but here also we probably have to do with essentially poetic symbols.

Heraclitus.—Perhaps a little more consideration should be given to Heraclitus of Ephesus (540–475 B.C.), who maintained that fire is the primordial substance. Here again the word fire was doubtless used as a symbol of the transitory, and this philosopher was apparently the first Western thinker to teach systematically that the senses are unreliable and that all things, even those which seem most permanent, are really 'moving pictures' made up by our minds from a series of constantly succeeding states. The firelight is apt to promote reflection, and one who attentively watches a candle or gas-flame may reach some important conclusions even if he knows very little about the mech-

[1] The reader will understand that most of these ancient dates may well be in error by a decade or more.

anism of combustion. A flame has many of the attributes of a rigid body such as form, position, temperature and inertia, yet it must be made up of constantly changing units whose entrance into it and departure from it no eye is able to perceive. Why then may not all our world be a 'flame picture' without permanency or enduring substance? Such conceptions are common enough in later thought. They did not serve to advance the knowledge of chemistry among the ancients.

Empedocles.—Empedocles (490–430 B.C.) combined the ideas of his predecessors and was apparently the first to speak of the 'four elements,' earth, air, fire and water. These he supposed to act upon each other under the influence of love and hate (attraction and repulsion). He seems to have associated a genuine chemical sense with these expressions, for he is said to have stated that flesh and blood are made up of equal quantities of the four elements, whereas bones are made of one-half fire, one-fourth earth, and one-fourth water. Such a statement in itself goes far to show the irresponsible rashness and self-confidence with which the philosophers of this age were wont to approach the discussion of things concerning which they knew nothing. Because coupled with the authority of great names this did untold harm in later centuries.

Democritus.—Democritus (470–560* B.C.) is frequently spoken of as the originator of the 'atomic theory' by persons who are not familiar with chemistry, and it is true that he did go far in formulating an attitude of mind which found expression in the real chemical atomic theory more than twenty centuries later.

Dialecticians have always been fond of discussing whether or not matter is infinitely divisible. To one type of mind natural phenomena seem most intelligible when matter is thought of as flowing and continuous so that, however often divided, each fragment is still in its turn divisible. Other minds find it more natural to assume that when subdivision has reached a certain point a particle is obtained which cannot be further divided, at least without losing the properties of the substance. Such persons are prone to account for the properties of the mass by the qualities of its component particles, and to explain changes

*These dates are especially uncertain.

in the mass by motion or interaction of the particles. Today we distinguish definitely between the so-called 'thermodynamic' and 'kinetic' schools of thought, and find them mutually helpful and supplementary. Democritus seems to have been the first of the kinetic school to definitely formulate its point of view, and so his fundamental conceptions have considerable interest for us. Like Leucippus, he thought of all things as made up of *atoms* to which he gave this name because they could not be further subdivided. He also stated that they were absolutely small, full, incompressible, without pores and homogeneous. Indeed he assigned them properties not unlike those of the 'mathematical point.' He permitted them to differ, however, in form, position and magnitude. Their properties could account in some measure for the properties of larger masses—thus, according to Democritus, water is a liquid because its atoms are smooth and round and can easily glide over each other. A solid like iron, on the other hand, must be made up of atoms which are hard and rough. Democritus indulged in many speculations in order to account for the first formation of bodies from the atoms and some of his other ideas are extremely interesting. He states, for example, that the function of respiration is to introduce new atoms into the body and to remove old ones. Like many of his contemporaries Democritus praised experiment as a valuable guide to knowledge, but as we have no record that he ever tried any experiments, we must consider what he said as merely a 'good resolution' which suffered the ordinary fate of such things. The spirit of the time was entirely against it.

Aristotle.—There is little record of what Socrates and Plato thought concerning natural phenomena. Aristotle (384–322 B.C.), however, gave no little attention to such matters and his ideas were destined to have great weight for many centuries. He distinguished between *matter* in a substance and what he called the *essence*. This difference may be the same distinction which later thinkers have drawn between the substance and the sum of its attributes. Without entering into a fruitless metaphysical discussion of this point, or raising what Huxley calls the "geometrical ghost" of a substance without attributes, we may say that what Aristotle called a substance he subdivided into essence and matter. The meaning of these terms may be made clearer

a particular period. The lists make no claim to completeness and, indee have been intentionally limited to works known to be reasonably accessib

One who desires to get into vital touch with the development of t science must, of course, study original papers in the pages of the journa and it is fortunate that so many of the less accessible of these have n been reprinted. The more casual reader, on the other hand, will frequent prefer a somewhat abridged résumé by a competent hand.

It is assumed for example, that only specialists will care to learn t views of the ancients on scientific matters by a study of their writings the original. The following two books, however, treat quite comprehensive the development of chemical science both among the ancients and in t mediaeval period:

HOEFER: *Histoire de la Chimie*, Paris, 1842.

KOPP: *Geschichte der Chemie*, Braunschweig, 1843.

The point of view of a writer in the forties was of course quite differe from our own, but this difficulty hardly makes itself felt in the study of t development of the science down to Lavoisier.

PLINY'S *Natural History* is available in an English translation by BOSTO and RILEY, 6 vols., Bohn's Classical Library, London, 1893.

CHAPTER II

CHEMISTRY IN THE MIDDLE AGES—ALCHEMY

Intellectual Decline in the Middle Ages.—The civilization of Greece in the time of Pericles or of Rome in the time of Augustus was intellectually upon an extremely high plane and the minds of educated men were at least as free from prejudice and superstition as they are today.

When, however, the Roman empire lost military and political power, and the governments of its provinces and vassal states became constantly more effete and corrupt, the moral and mental tone of the community was lowered at the same time, and with the decay of manners crept in mental indolence and inefficiency. Finally, when the weakness of the empire invited the cupidity of the barbarian hordes their conquests smothered intellectual life altogether, and we can record little constructive scientific work or scientific thinking until the revival of learning in the fifteenth century. The unsettled conditions of the times rendered a life devoted to study impossible except in the monasteries, and even here intellectual freedom was so hampered by an inflexible dogmatic theology that the desire for intellectual activity could find expression in little save the copying of manuscripts or the hair-splitting futilities of the scholastic philosophy.

During this period the sciences fared even worse than art and letters, because, with the possible exception of mathematics and astronomy, they had made a much poorer start. In fact, there was little real progress throughout the entire fifteen centuries. We have, however, to take account of new conditions and a new point of view.

Alchemy.—The slow progress of science among the ancients was due to the divorce of theory and practice. Those who did the work and those who did the thinking were entirely out of touch. In the Middle Ages, on the other hand, the theories were indeed evolved by the same men who did the experimenting, but these were frequently persons of inferior mentality, whose work was usually poor and whose thinking was apt to be slovenly when it was not actually dishonest.

From the chemist's standpoint, the most important intellectual symptom of this time is the rise and spread of Alchemy. By this we commonly understand the pretended art of changing the baser metals into gold. Such a definition suggests quackery and self-seeking, and there is no question but that many alchemists were no better than common cheats, especially in later times. We should make a grievous mistake, however, if we condemned all so sweepingly, or imagined that a vulgar cupidity was the only impulse which started men upon the quest for the philosopher's stone. The literature of alchemy is full of turgid rhetoric and mystic symbolism but through it all runs the idea that the change of the base into the noble has not only a chemical but a moral significance, and that he who discovers the "stone of the ancient sages" will also reap another reward in the enrichment of his mind and the elevation of his character. Furthermore we all remember how our first acquaintance with chemical transformations ministered to our taste for the marvelous which after all is not so far from a true scientific interest. In short, since alchemy was the only chemistry in those days, we can readily see that men pursued it for many motives, then as now. Liebig has well said: "To one man science is a sacred goddess to whose service he is happy to devote his life; to another she is a cow who provides him with butter." So it was with alchemy.

Relation of Alchemy to Science.—Perfect clearness is also necessary on another point which is sometimes misunderstood. We must remember that there was nothing inherently absurd in the problem which the alchemists set themselves. It is the essential nature of chemical change that one substance with certain properties disappears while another with different properties takes its place, and there was nothing in the knowledge of the times from which one had the right to conclude that it was any more impossible to obtain gold from lead, than to obtain lead itself from litharge or mercury from cinnabar. In fact, the recent preparation of helium from radium puts the logic entirely on the side of the alchemists. Chemistry is an experimental science, and the only way to find out whether it is practicable to get gold from lead is to try it. We owe it to the alchemists to acknowledge that they did try it, with crude means, it is true, but with endless patience and much needless repetition. The pity of it is that all their efforts were so wasted by secrecy that they could show few results of value in a whole thousand years.

Origin of Alchemy.—The origin of alchemy is obscure. The first authentic literature belongs to the fourth century and is associated with Alexandria. Zosimus of Panopolis seems to have been the first of these voluminous writers. Little of his work is preserved but what we have has the essential characteristics found in so large measure in all subsequent alchemistic writings. They are characterized by a bombastic mysticism, apparently written in a kind of religious ecstacy, and while they contain numerous chemical recipes these are couched in unintelligible language and alternated with high-sounding invocations. Nevertheless, in all this meaningless jargon we are sometimes refreshed by phrases which show genuine insight and by flashes of real humor.

The traditional view among historians has been that these earliest books represent materials which had been handed down orally for some time by the Egyptian priests, and may have had something to do with their religious ritual. The eminent French chemist Berthelot was of a different opinion. He gave much study to the so-called "Leiden Papyrus" which was originally found in a tomb in Thebes. On translation this proved to be a book of workshop receipts, in which among other things there are directions for so mixing and coloring metals as to imitate gold for the manufacture of cheap jewelry. There is no thought here of transmutation, but Berthelot believed that a goldsmith who had become expert in such arts might, in the absence of all chemical standards, deceive himself into the belief that he had actually effected a transmutation. Such a discovery would be highly profitable, and the finder would be apt to pass on the secret only in terms which could not be easily understood. In this way, as others guessed, experimented, copied, and wrote directions in their turn, a literature like that of alchemy might spring up.

Alchemistic Traditions.—Whatever the facts, if we were to accept the traditions of the alchemists themselves concerning the origin of their art we should come to very different conclusions. Some of their accounts ascribed it to the days of communion between men and angels in the period before the Flood, and numbered most of the patriarchs, including Adam himself, in the ranks of the alchemists. This, of course, was meant to play upon the human tendency, stronger then than now, to value everything in proportion to its antiquity. The same influence led some alchemists to forego fame and ascribe their own writings to the ancient philosophers, Democritus being a favorite.

Alchemistic tradition also has much to say of a certain Hermes Trismegistos (probably connected in some way with Thot the Egyptian god of wisdom, who also was represented as carrying a rod entwined by serpents). Concerning the dates and places of residence of this Hermes there is no agreement, but one tradition has it that he inscribed upon an emerald the most essential secrets of alchemy and presented this jewel to Sarah, the wife of Abraham. After many vicissitudes the stone was lost, but the traditional wording of the inscription has come down to us, and is worthy of a place here because alchemists in general seem to have taken Hermes as a model in style and clearness. We also pay tribute to his name when we speak of sealing a vessel hermetically.

The Emerald Tablet

It is true and without falsehood, certain and most true, that which is above is even as that which is beneath. And that which is beneath is even as that which is above, for accomplishing the miracles of one thing.

And as all things were from one by the meditation of one, so all things were born from this one thing by adoption.

Its Father is the Sun, its Mother is the Moon. The wind carried it in its belly. Its nurse is the earth. This is the father of all the knowledge of the whole world. Its virtue is unimpaired if it should be turned toward the earth.

You will separate the earth from the fire, the subtle from the compact, gently, with great skill. It ascends from earth to Heaven, then descends again to Earth and receives the force of those above and those below.

Thus you will possess the glory of the whole world and all obscurity will flee from you.

This is the strong strength of all strength, because it will overcome every subtle thing, and penetrate every solid.

So the World was created.

There will be wonderful adaptations of which this is the mode.

Therefore am I called Hermes Trismegistos having three parts of the philosophy of the whole world.

It is finished, what I have said concerning the operation of the Sun.

Fundamental Ideas of the Alchemists.—Language of this kind is hardly capable of expressing any rational idea but, judging by their clearer writings, the alchemists seem to have held that there was a certain *materia prima* which was present in all things though always contaminated by impurities. These they hoped to remove by processes of purification, especially by fire (calcination, sublimation or distillation), and in this way they expected to obtain the "essence" or "tincture," which was apparently identical with the *philosopher's stone.* Once obtained this was expected to work wonders of many kinds. It would change the baser metals to gold by contact, it would heal all diseases and even regenerate the character of the fortunate discoverer. In addition to the *materia prima,* which may have been originally derived from the "quintessence" of Aristotle, the alchemists generally recognized the other four elements which he had accepted, although they frequently added or substituted others of their own, particularly mercury, sulphur and salt. Here, however, they were always careful to point out that they did not refer to the substances commonly known under these names but rather to the mercury, sulphur and salt "of the sages." Mercury seems to have stood for the metallic and also the volatile character in general, and the mechanism of transmutation was frequently referred to as "fixation of the mercury." In the same way sulphur represented the property of combustibility, whereas salt stood for the salty or earthy properties, notably resistance toward fire. It must be remembered, however, that all these terms were constantly used in the most reckless and inconsistent way, as the following selection from Basilius Valentinus abundantly shows:

"That there can be no perfect generation or resuscitation without the coöperation of the four elements, you may see from the fact that when Adam had been formed by the Creator out of earth, there was no life in him until God breathed into him a living spirit, then the earth was quickened into motion. In the earth was the salt, that is the body; the air that was breathed into it was mercury or the spirit, and this air imparted to him a gentle and temperate heat which was sulphur or fire. Then Adam moved, and by his power of motion showed that there had been infused into him a life-giving spirit. For as there is no fire without air so neither is there any air without fire.[1] Water was incorporated with the earth. Thus living man is a harmonious mixture of the four elements; and Adam was generated out of earth, water, air and fire; out of soul, spirit and body; out of mercury, sulphur and salt."

The recipes and directions in alchemistic books laid great weight upon the phases of the moon, the position of the planets and the utterance of appropriate incantations, and this need not surprise us when we recall that in these early times no means were available by which such essential conditions as temperature and pressure could be regulated. Furthermore, the materials employed could seldom have been pure or even uniform. When we add that there were no analytical methods by which materials could be tested, we see that the results of experiments must frequently have seemed utterly capricious, and it was natural that weight should be laid upon trifling and irrelevant circumstances.

The planets caused especial trouble. From early times the 'seven planets' had been associated with the 'seven metals' and in writing the same astronomical signs were commonly employed for both. The sun stood for gold, the moon for silver,[2] Jupiter for tin, Saturn for lead, Mars for iron, Venus for copper, and Mercury for the metal of the same name. It followed naturally that when a metal was to be acted upon chemically in a certain way its 'patron planet' must be rightly situated.

Practical Achievements of the Alchemists.—In spite of these handicaps the world gradually did accumulate scientific information. The tendency to heat, distil and combine all obtainable substances in order to obtain the philosopher's stone had the practical result that many important reactions were observed

[1] The writer had evidently made the important observation that there is "no fire without air" but the last half of the sentence shows how much he preferred a striking antithesis to any fact of observation.

[2] The term "lunar caustic" for nitrate of silver is an inheritance from those early times.

and many important compounds prepared. Unfortunately the alchemists were so unwilling to use intelligible language in the description of their discoveries that most of these remained unfruitful.

No History of Alchemy Possible.—For the same reason no real history of the alchemistic period is possible in spite of the fact that the literature is surprisingly voluminous. In these writings fragments of experimental detail here and there show plainly enough that the adepts of the fifteenth century had more chemical knowledge than those of the fifth, but when we look for theories or a fundamental point of view all are so contradictory and obscure that no rational progress can be traced,

> *"Denn ein volkommner Widerspruch*
> *Bleibt gleich geheimnisvoll für kluge wie für Thoren."*

Prominent Alchemists.—It will be appropriate, however, to mention a few of the prominent names commonly associated with alchemy. After the Alexandrian school the most prominent alchemist is an Arab of the eighth century. His full name is very long and is variously quoted, so that Western writers by common consent now call him simply Geber. The place of his residence is uncertain but his numerous works brought him great fame in the Middle Ages. Some of these are still extant in Arabic while certain Latin treatises purporting to be translations of Geber are preserved in European libraries. The latter books date from about the thirteenth century and Berthelot, after comparing them with the Arabic writings of Geber, came to the conclusion that they were the work of another hand, and had been attributed to Geber in order to enhance their prestige. It is interesting to note that Berthelot found the Latin works superior from the chemical point of view.

Other noted men claimed as alchemists are Albertus Magnus (1206–1280), Roger Bacon (1214–1294), and Raymund Lullus (1235–1315). Extensive treatises on alchemy are commonly ascribed to these men, but the best modern opinion holds that these are spurious. Doubtless they believed in the possibility of transmutation because that represented the 'consensus of scientific opinion' of their day, but their actual participation in alchemistic work cannot be proved, and is inconsistent with most of what is known of their other activities, except perhaps in the case of Bacon.

Lullus was a poet and philosopher who finally lost his life on a missionary expedition, being stoned to death by the natives. Albertus Magnus has been sainted. He was Bishop of Regensburg in 1260, was a friend of Thomas Aquinas and is mentioned by Dante. Roger Bacon, who won from his contemporaries the appellation *Doctor Mirabilis*, was an inventive genius much in advance of his time who added materially to knowledge by his mathematical studies and experiments. A great number of practical inventions are commonly ascribed to him, though in many cases the proofs are lacking. Among his writings are found a recipe for making gunpowder, directions for the construction of a telescope and a study of the rainbow which is said to be extremely good.

A more genuine alchemist was Basilius Valentinus, author of the *Triumphal Chariot of Antimony* and several other less-known works. He is reputed to have been a Benedictine monk and to have lived in the latter part of the fifteenth century, but practically nothing is really known of his life. His writings are clearer than those of his predecessors so that many of his experiments can be repeated and verified. He did a real service in characterizing the metal antimony and many of its more important compounds.

Decay of Alchemy.—After the fifteenth century as chemistry became gradually more scientific, alchemy in the narrow sense tended to die out and acquired an ill repute through the character of its devotees. The more brilliant of these secured positions at the courts of petty princes where they made a precarious living by playing upon the avarice of their patrons, while the less fortunate ones practised similar frauds in a humbler sphere. Finally conditions must have become extremely bad, for the story goes that Frederick of Würzburg maintained a special gallows which he employed solely for hanging alchemists. The careers of these men were sometimes extremely romantic, but with this the history of chemistry has, of course, nothing to do.

Literature

In addition to Hoefer and Kopp the following books by Berthelot are of interest:

Les Origines de l'Alchimie, 1883.
Collections des Anciens Alchimistes Grecs, 1887–8.
La Chimie au Moyen Âge, 1893.

A. E. Waite has also done a distinct service by translating into English and publishing a number of old alchemistic writings. Among these may be mentioned *The Triumphal Chariot of Antimony* by Basilius Valentinus, *The New Pearl of Great Price* by Bonus of Ferrara, and an extremely interesting old collection called *The Hermetic Museum*. These are all published by Elliott of London.

CHAPTER III

CHEMISTRY IN THE RENAISSANCE

Attention has already been called to the eclipse of free intellectual life which accompanied the downfall of the Roman Empire. During the Middle Ages there was little mental activity save that which was under the direct protection of the Church, and here everything naturally centered upon the exposition of the Scriptures and the writings of the Fathers, indeed scarce any profane literature was looked upon with favor except the writings of Aristotle who, strangely enough, enjoyed an authority not inferior to that of the saints and martyrs.

The Revival of Learning.—In the fourteenth century there began a new intellectual movement which found expression everywhere as a reaction against authority and an assertion of the rights of the individual to think and act for himself. First, came a revival of interest on the part of Italian scholars in the secular literature of ancient Greece and Rome which brought a realization of the greater intellectual freedom of ancient times and an ardent desire to imitate it in letters and in art. Momentum was given to the movement by the fall of Constantinople in 1453 which scattered the scholars of the Eastern Empire throughout the West, while the invention of printing which came at about the same time made possible a hitherto undreamed-of multiplication of books and gave a new impulse to literary effort. This was also the age of great discoveries. The voyages of Columbus and Vasco da Gama opened new fields to exploration and many thousands in the spirit of Cortez, Pizzarro and Magellan sought fortune and adventure in the New World. At the same time a revolt against religious authority was going on in Germany. Luther posted his theses upon the church door at Wittenberg in 1517, and about the same time even the sciences showed some signs of an awakening. Of these astronomy had fared best in ancient times because its study was so closely

BASILIUS VALENTINUS

(*Facing page 18*)

from Hippocrates, Galen, or any one else, but by experience, the great teacher, and by labor, have I composed them. Accordingly, if I wish to prove anything, experiment and reason for me take the place of authorities. Wherefore, most excellent readers, if any one is delighted with the mysteries of this Apollonian art, if any one lives and desires it, if any one longs in a brief space of time to acquire this whole branch of learning, let him forthwith betake himself unto us at Basle and he will obtain to far greater things than I can describe in a few words. The ancients gave wrong names to almost all the diseases; hence no doctors, or at least very few, at the present day, are fortunate enough to know exactly diseases, their causes, and critical days. Let these proofs be sufficient, notwithstanding their obscurity. I do not permit you to rashly judge them before you have heard Theophrastus. Farewell. Look favorably on this attempt at the restoration of medicine."

And again:

"In the meantime, I extol and adorn, with the eulogium rightly due to them, the Spagyric physicians. These do not give themselves up to ease and idleness, strutting about with a haughty gait, dressed in silk, with rings ostentatiously displayed on their fingers, or silver poignards fixed on their loins, and sleek gloves on their hands. But they devote themselves diligently to their labors, sweating whole nights and days over fiery furnaces. These do not kill the time with empty talk, but find their delight in their laboratory. They are clad in leathern garments, and wear a girdle to wipe their hands upon. They put their fingers among the coals, the lute, and the dung, not into gold rings. Like blacksmiths and coal merchants, they are sooty and dirty, and do not look proudly with sleek countenance. In presence of the sick they do not chatter and vaunt their own medicines. They perceive that the work should glorify the workman, not the workman the work, and that fine words go a very little way toward curing sick folks. Passing by all these vanities, therefore, they rejoice to be occupied at the fire and to learn the steps of alchemical knowledge. Of this class are: Distillation, Resolution, Putrefaction, Extraction, Calcination, Reverberation, Sublimation, Fixation, Separation, Reduction, Coagulation, Tincture and the like."

The fundamental idea of Paracelsus seems to have been that life is essentially a chemical process. If, then, man is a chemical compound (as the theories of the day would seem to demand) of mercury, sulphur and salt, then good health must be the sign that the elements are mingled in the correct proportions, but illness shows that one or more of these 'elements' is deficient. The logical treatment, therefore, is to dose the patient with that which he lacks in some form suitable for assimilation. Such considerations induced Paracelsus to abandon the herbs and extracts chiefly used by the physicians of his time and to prescribe inorganic salts. Indeed, mercury and its compounds owe their present prominence in the pharmacopœia originally to him.

As we read of these theories we must ask ourselves with some dismay what could have been the state of therapeutics in the fifteenth century if *this* was an improvement! As a matter of fact the medical profession was long divided between the old doctrines and the new, the disciples of Paracelsus idolizing him and pointing with pride to his marvelous cures while his enemies denounced him as a quack. With these controversies and the other interesting physiological ideas of Paracelsus we are not concerned. His service to chemistry consisted essentially in this, that he induced the alchemists to give up the search for gold and to devote their chemical skill to the preparation of remedies, while at the same time he compelled the physicians to learn a little chemistry. This did good in both directions. The gain to chemistry was that the medical profession included then as always educated men, whose mental power far surpassed that of the alchemists of the day. We must not leave Paracelsus without recalling his magnificent motto, "Let him not belong to another who may be his own."

Agricola.—There could have been no greater contrast to Paracelsus than his contemporary Georg Agricola (1490–1555). The latter spent most of his life in Joachimsthal and Chemnitz, and though a physician he is best remembered by his work in metallurgy. Without interest in the tumultuous controversies of his time, he devoted himself to making valuable observations in his chosen field and recording them with accuracy and clearness. His real services are, therefore, greater than those of Paracelsus, but just at this time a fiery controversialist like the latter was needed to set men thinking. It is a splendid tribute to Agricola that his great work *De Re Metallica* served as a valued handbook in metallurgy until comparatively recent times. The author's mental attitude, however, found little imitation in his own day, and the controversy between the Paracelsian and Antiparacelsian schools of medicine raged on, men like Torquet de Mayerne and Adrian de Mynsicht supporting the views of Paracelsus, while Andreas Libavius was among the most prominent of those who opposed the new doctrine. In this controversy argument and denunciation so far exceeded experiment that not as much was won for the science as ought to have been the case. Indeed the Paracelsians dared not submit some of their views to the test of experiment because the fundamental doctrine of their master—that man was composed of mercury, sulphur and salt—could not be verified by decomposing him again into these substances.

Van Helmont.—Very considerable interest attaches to the work of the Belgian physician, Jan Baptista Van Helmont (1577–1644). Van Helmont must have possessed a peculiar personality, for with an innate taste for the mysterious and occult he still combined the capacity for accurate observation and clear thinking. In consequence his writings contain strange contradictions. He not only believed in the transmutation of metals but claims that he had himself accomplished this, and his attitude toward several other alchemistic traditions is equally credulous. In contrast it is of interest to trace one of the experiments which led him to reject the elements of Paracelsus and adopt water as the primordial substance. Finding a small willow weighing only five pounds, he planted this in two hundred pounds of earth and watered it regularly for five years. At the end of this time he removed the earth from the roots and found its weight unchanged, while the willow now weighed a hundred and sixty-nine pounds. From this he concluded that at least one hundred and sixty-four pounds must represent water, since the earth had not changed in weight and the willow had received no other nourishment. We must consider this an exceptionally good scientific investigation for the times. We are tempted to smile at the conclusion but the fact is that Van Helmont had no means of observing the assimilation of carbon dioxide by the plant. There is nearly twenty times as much argon in the atmosphere as carbon dioxide, and since the former gas remained undiscovered until the closing years of the nineteenth century, we must forgive Van Helmont for underestimating the importance of the latter. It is interesting, however, that this oversight should have been made by the very man who discovered carbon dioxide, for it was Van Helmont who first recognized that a gas which did not support combustion was formed when wood is burned, and that the same substance is produced by the action of acids upon limestone and during the process of fermentation. Indeed he is the first writer to use the word "gas," and he distinguishes such substances from vapors as less easily converted to liquids. In medicine, Van Helmont considered most physiological processes as fermentations and dwelt less than his predecessors upon hypothetical elements and more upon substances actually found in the body. This led him to classify diseases as acid and alkaline and to treat them by neutralization. As successors of Van Helmont may be mentioned François le Boe Sylvius (1614–1672) and Otto Tacchenius who died in 1675. Their contributions to chemistry and medicine were along similar lines.

Glauber.—No account of this period would be complete without mention of Johann Rudolph Glauber (1604–1668) who perhaps came nearer than any of his predecessors to being a chemical engineer. Glauber lived chiefly by the sale of secret medical preparations, and his writings, which are delightfully quaint, abound in all sorts of alchemistic superstitions. He was, nevertheless, a shrewd observer, and quite original in his thinking. He wrote a number of books, and one is noteworthy because it is essentially a treatise upon political economy. It is entitled *Teutschlands Wohlfahrt* and points out how Germany may develop its own resources, especially along chemical lines, and so become independent of other countries. The author's name is perpetuated for us in 'Glauber salt,' a designation still retained for crystalline sodium sulphate. It is described in his *Miraculum Mundi* where he calls it *sal mirabile* not only on account of its value as a remedy but, among other things, for its property of dissolving carbon. We now know of course that when carbon is fused with sodium sulphate no real solution takes place. The sulphate is reduced to sulphide while carbon dioxide escapes. To Glauber, however, the disappearance of the carbon was evidence that it had been dissolved. He showed how sodium sulphate could be prepared from common salt by the action of sulphuric acid and he used similar methods for the preparation of other acids, notably nitric acid. He also interpreted correctly many cases of metathesis, a class of reaction not previously well understood. He also described the preparation of acetic acid by the distillation of wood and other processes which have since been developed on the large scale. At the time of Glauber progress was being made in many of the industries associated with chemistry. Pallissy had made considerable improvements, not only in pottery but also in rational agriculture, Venetian glass-makers were doing some of their most skilful work, Agricola had laid the foundations of metallurgy and made valuable beginnings in assaying. The art of dyeing was also becoming improved and more systematic, the first handbook devoted to this subject appearing about 1540.

Literature

There is a life of Paracelsus in English by HARTMAN, London, 1887. See also WAITE, *Hermetic and Alchemical Writings of Paracelsus*, 2 vols., London, 1894.

Original editions of Glauber's books both in Latin and German are still extant though rare.

Robert Boyle

John Mayow

CHAPTER IV

BOYLE AND HIS CONTEMPORARIES—THE PHLOGISTON THEORY

Chemistry made but slow progress even in the seventeenth century, but from this time on we shall meet with men who were willing to pursue the study in the same spirit in which Galileo, Huyghens and Keppler were devoting themselves to astronomy and physics.

Boyle.—First among these comes Robert Boyle (1627–1691) the most broad-minded and widely cultivated man who had yet interested himself in chemistry. Boyle was a younger son of the Earl of Cork who sent him to Eton at the age of eight. Three years later he went to the Continent for the completion of his studies and remained till 1644, when he returned to England and took up his residence at Stalbridge Manor in Dorset. Here Boyle became associated with a club of progressive men interested in science who, because they had no fixed place of meeting, called themselves "The Invisible College." In 1644 Boyle removed to Oxford and in 1680 to London. The organization was unpopular at first, as all associations are apt to be whose members 'seek new things,' but good fortune came from an unexpected quarter when Charles II saw fit to dabble in science. Experimentation grew fashionable at Court and the Invisible College, under the king's favor, became the Royal Society, an institution destined to be an important agency in the advancement of science from that day to this. Boyle was prominent in the councils of the society until his death.

Work in Pneumatics.—He is best remembered by his work on pneumatics. To appreciate this we must recall that previous to his time little progress had been made toward explaining the action of so simple a mechanism as the suction pump. It was commonly stated that 'Nature abhorred a vacuum,' and accordingly when air was removed by the piston, water must go in to take its place. Those who tried to raise water more than thirty-four feet by this means, however, soon found that under these circumstances the "abhorrence" was not sufficient to produce practical results. At last it occurred to Torricelli that it was the pressure of the air upon the surface of the water which forced the latter into the tube. If this were true it stood to reason that the atmosphere would balance a much shorter column of mercury than of water. Torricelli tried this in 1643 and so invented the barometer. A little later Pascal observed that when a barometer was carried to a height the mercury fell, another confirmation of Torricelli's views. Boyle attacked the problem from a somewhat different angle. Otto von Guericke, the inventor of the air-pump, had published an account of his discovery in 1654. Reading this, Boyle decided to construct a new and superior pump which he completed in 1659. With this instrument he tried many experiments. Among others he placed a barometer under the receiver, and when the air was removed by the pump he had the pleasure of observing that the mercury continually fell, proving conclusively that it was the pressure of the air which supported the column. In 1660 an account of these experiments was published in a treatise *On the Spring of the Air*. The book was attacked by one Franciscus Linus who explained the barometer in his own fashion, maintaining that the mercury column was sustained by an invisible internal cord. This explanation was characterized by Boyle as harder to understand than the facts were without it, which is in itself no mean test for a hypothesis. In his reply to Linus, Boyle states his famous law that the volume of a gas varies inversely as the pressure, and he describes the experiment by which he established this, air being confined in the closed arm of a U-tube while it was subjected to varying pressures by pouring mercury into the other arm.

In addition to this important discovery Boyle did much valuable work in fields strictly chemical. He was the first to use the term "chemical analysis" in its modern significance. In fact, he did more to systematize the various qualitative tests then in use than any of his predecessors. His 'complete works' are a formidable production for he was a voluminous and prolix writer, who delighted in the Platonic dialogue. One of his works which did a real service was *The Sceptical Chemist*, published in 1661. In this he attacked the 'elements' of the alchemists and defined the term in the modern sense as something which has not been decomposed.

Mayow and Hales.—Among Boyle's writings there is a paper upon the function of the air in combustion but in this he was not so fortunate in his conclusions as his younger contemporary, John Mayow (1645–1679), who recognized that the air contains a substance which unites with metals when they are calcined, that it changes venous to arterial blood, and that it occurs in saltpeter. For this reason he called it *spiritus nitro-aerius.* It seems possible that if Mayow had lived longer he might have discovered oxygen and so given the world a truer conception of the nature of combustion than it was destined to have for years to come. Somewhat later than Boyle and Mayow lived Stephen Hales (1677–1761), a clergyman who took interest in the study of gases and devised many clever means for their manipulation. Mayow and Hales may be regarded as the direct forerunners of Priestley.

Kunkel and Becher.—The mental attitude shown by Boyle and his associates in England represented, so far as chemistry was concerned, a more advanced position than had yet been attained upon the Continent. Here Boyle's most prominent contemporaries were Johann Kunkel (1630–1703) and Johann Joachim Becher (1635-1682). The former was a chemist at various courts and an expert in the manufacture of glass. His *Ars Vitraria* is a comprehensive treatise on this subject. After many vicissitudes Kunkel finally found favor and a title at the court of Charles XI of Sweden. Becher was an uneasy spirit who divided his time between teaching, chemical theorizing, and the promotion of various socialistic and financial schemes which frequently involved industrial applications of chemistry; thus he was the first to take out a patent for distilling coal, suggesting that the tar might serve for the preservation of cordage while the gases would be suitable for smelting since they gave "a flame ten feet long." The world, however, was not yet ready for a coal-tar industry so this venture and many others like it came to nothing. Becher lost the friendship of those who had put money into his schemes, and more than once he had to flee the country, though no charges of personal dishonesty are recorded against him. Becher's writings on chemistry had no little vogue. They were for the most part visionary and alchemistic in spirit but in one point they were destined to have a great influence upon the thought of the eighteenth century. Becher adopted as elements three "earths"—the "mercurial," the "vitrifiable" and the "combustible." It was this last, the *terra pinguis*, which was destined in the hands of Stahl to become the foundation of the great phlogiston theory.

Stahl.—Georg Ernst Stahl was born in Anspach in 1660. He graduated from the University of Jena in 1683 and became physician to the Duke of Weimar in 1687. In 1694 he was made professor of medicine at Halle and in 1716 became physician to the king of Prussia. He died in 1734.

The Phlogiston Theory.—Stahl is best known as the founder of a system of chemical philosophy—the first of those comprehensive theories which have successively since his day dominated chemical thought. The subject is interesting, therefore, not only on its own account but for what it can teach us concerning dominant theories in general.

The fundamental idea in the phlogiston theory was that all combustible substances possessed one component in common which escaped in the act of burning. For this Stahl expressed his indebtedness to Becher, but the idea itself was really much older since the alchemists used the term "sulphur" in much the same sense. The alchemists, however, were content to acknowledge this as a dogma and disregarded it at their pleasure, while Stahl treated his phlogiston as a definite chemical component, and used it as a guide in everyday laboratory practice.

Reactions Explained by Phlogiston.—The theory was clearly based upon the common-sense observation of that kind of combustion with which everybody is most familiar. When a piece of wood burns we seem to see flames issue from it at every pore and pass upward. The wood blackens, cracks open, and when the flames are gone, the fragments glow for a few moments and then crumble to ashes. What more natural conclusion than that a fire-substance, *phlogiston*, has escaped while the ashes are left? It would follow from this that the wood is a compound of phlogiston and the ash. This is capable of wide expansion. If phosphorus be burned instead of wood, the 'ash' is white and

Georg Ernst Stahl
1660–1734

(*Facing page 28*)

HENRY CAVENDISH
1731-1810

CHAPTER V

THE LATER PHLOGISTIANS—THE DISCOVERY OF OXYGEN

Black on Magnesia Alba.—With Joseph Black (1728–1799) we ome to the eminent group of distinguished chemists whose work ontributed so largely to the overthrow of the phlogiston theory, nd it is especially interesting to see how they themselves almost 'ithout exception remained blind to this, its most important gnificance. Black was long professor in Glasgow, and made)me important discoveries in physics, developing independently ıe idea of specific heat and of latent heat, though his work was ot formally published. He is best remembered by chemists for is work on *magnesia alba* which he presented for the doctor's egree in 1754. In this investigation he took up the study of hat we now call magnesium carbonate as a new substance which e desired to characterize, and he proceeded to try some experiıents upon it from which he was able to draw important concluons. In the interest of compactness we may sum up Black's esults in a series of propositions:

I. *Magnesia alba* when strongly heated loses about half its 'eight and yields a new substance *magnesia usta* (magnesium xide).

II. With vitriolic acid *magnesia alba* yields epsom salt (magesium sulphate) with effervescence.

III. *Magnesia usta* when similarly treated yields epsom salt ithout effervescence.

IV. In a solution of epsom salt, mild alkali (potassium caronate) precipitates *magnesia alba* and the solution on evaporaion yields vitriolated tartar (potassium sulphate).

V. Mild alkali effervesces with acids while caustic does not.

VI. Mild alkali is made caustic by addition of *magnesia sta*.

When arranged in this form it is particularly easy to see what handicap upon the chemists of that time was the use of a nomenclature necessarily incapable of expressing chemical relationships, and how impossible it then was to know whether all substances in a reaction were accounted for or not. Nevertheless, Black interpreted his results with perfect accuracy. From II and III he concluded that the difference between *magnesia alba* and *magnesia usta* was the gas ("fixed air") liberated from the former by acids, and that it was the expulsion of the same gas which accounted for the loss of weight when magnesia alba was heated (I). II and IV showed that *magnesia alba* could be regenerated from *magnesia usta* by the aid of mild alkali, hence the latter must contain fixed air which it surrenders in the reaction. This is further confirmed by V which shows that mild alkali differs from caustic by its content of fixed air. Finally VI completes the caustifying of the alkali by the action of magnesia. Black saw at once that these reactions were analogous to those involved in the ancient method of preparing caustic alkali from quicklime. He accordingly repeated his experiments using limestone instead of *magnesia alba* and so reached the correct conclusion that the 'burning' of lime consists essentially in the expulsion of fixed air.

Such a result was utterly opposed to the explanations hitherto current. According to the latter, when lime was heated in the kiln phlogiston entered into it making it fiery or caustic. Later when the quicklime was treated with mild alkali another transfer of phlogiston occurred and the latter became caustic in its turn.

It is what might have been expected from his clear habits of thought that in later years, when Lavoisier had once shown the way, Black was among the first to adopt the new views.

Cavendish.—England is conspicuous for the number of its men of wealth and family who have devoted their lives to science. Boyle was a prominent example and we find another in Henry Cavendish (1731–1810), a nephew of the third duke of Devonshire. Cavendish was of an eccentric turn, and countless stories are told of his strange habits, his shyness and his aversion to women. He lived as a recluse and devoted practically his entire time to research, and although in middle life he inherited a fortune which made him one of the richest men in England, this had no influence upon his regular and frugal habits. As might be expected in such a character his work was done with little thought of

fame, and much of the best of it remained entirely unknown till long after his death. Cavendish was the first to make a thorough study of hydrogen and he gave it the name of "inflammable air" in a paper published in 1766. The evolution of a combustible gas when a metal is dissolved in acids was observed much earlier. We are reasonably sure that it was known to Paracelsus and Van Helmont, and we know that it was isolated by Boyle. Cavendish identified hydrogen with phlogiston, and this was entirely in the spirit of current views, for if a metal is a compound of a base with phlogiston then when the base unites with an acid to form a salt phlogiston must escape.

About 1783, after the discovery of oxygen, Cavendish combined this gas with hydrogen by means of the electric spark, and so established the composition of water. In 1785 while conducting experiments of this kind he noticed that oxygen and nitrogen when sparked in this way over water yielded nitric acid, and applied the idea to the complete absorption of atmospheric nitrogen. He always found, however, a small inert residue whose volume could not be further reduced and which he estimated at about $\frac{1}{120}$ of the whole. It is interesting to note that in spite of this valuable clue so faithfully recorded, argon and the other rare gases of the atmosphere remained undiscovered for more than a hundred years.

It is now hard to see how Cavendish could have accounted for his results in terms of the phlogiston theory, but he did so on the assumption of Priestley that oxygen was "dephlogisticated air," that portion of it, namely, which unites with phlogiston on combustion. He was not ignorant of the work of Lavoisier, and acknowledged frankly that the latter's views would explain the results of his experiments "nearly as well," but after weighing both opinions he clung to the old for what now seems a curious reason. He writes:

"There is one circumstance also, which though it may appear to many not to have much force, I own has some weight with me; it is, that as plants seem to draw their nourishment almost entirely from water and fixed and phlogisticated air, and are restored back to those substances by burning, it seems reasonable to conclude, that notwithstanding their infinite variety they consist almost entirely of various combinations of water and fixed and phlogisticated air, united according

20 *Fundamenta Chymiæ*

ſchieſst, vor ſich trocken bleibt, einen ſtumpff weichlich-ſüſſerlichen Geſchmack hat, der gleichſam eine Kälte verurſachet, ſo ſchieſſt dieſs hingegen nimmermehr an zu Cryſtallen, ſchmeckt ſcharff & valde quaſi urit os, zieht aus der Lufft Waſſer an ſic. & hoc eſt quod nitrum fixum vocant.

Ubi ego cauſam hanc oſtendi, quod uno momento nitri mixtio diſſolvatur, adeo, ut, qui tam cito vel milleſimam partem hujus portionis nitri reparare poſſet, dives facile fieri poſſet. Denn es gehen in dieſer Diſſolution des nitri, indem ſich die Mixtur entzündet, variæ nitri portiones in forma flammæ weg, und machen einen Rauch welchen einige künſtlich, applicato ſcil. ſuper crucibulum recipiente, fangen; nemlich ſie thun dieſe Mixtur in ein glüend Geſchirr nach und und nach, und der Dunſt, der ſonſten in die Lufft gehet, wird in den vorgelegten Recipienten gefangen, da er ſich in einen Liquorem condenſiret, qui vero plane alienus eſt à Spiritu nitri indole: denn wenn man ſonſten die portionem nitri fixam auf eine andere Weiſe von der portione volatili & aquea ſcheidet, bekömmt man auch einen Dunſt der ſich von der fixa portione ſondert, und endlich gleichſam zu Waſſer wird: dieſes iſt aber ein ſehr ſtarker und penetranter ſpiritus, deſſen auch nur ein Tropff in kurzer Zeit ein ziemlich Loch in die Haut friſſet, ja, ſo man ein ſubtil Oel ihm zugieſſt, erhitzt er ſich und feuert, als wenn er ſich entzünden wolte, welches auch würcklich geſchicht (daſs er nemlich eine Flamme wird) wenn man ihm ein gewiſſes Salz zuſetzet.

Ab hac ſpiritus hujus nitri indole plane aliena eſt, hæc de qua loquimur aquea nitri portio (quæ nempe à fixatione nitri prodit) utpote quæ plane eſt inſipida, & odoris non niſi fuliginoſi &c. In dieſer Operation (uti monui) die recht in momento geſchicht, kan man ſehen, wie die intima mixtio im Moment zerſtöret wird, alſo daſs nicht einmal ein veſtigium nitri zu ſpüren, und ſolte einer nicht meynen daſs es nitrum geweſen; ſal autem illud alcalicum, quod remanet, proprie ad nitri eſſentiam non pertinet. Es wäre gut, wanns einer auch wieder ſo geſchwinde reſtituiren und aus der Lufft, in welche es ſo geſchwinde weggeflogen, wieder herholen könnte. Chymicis præterea, non vulgo (utpote cui nota quidem eſt res, cauſæ tamen incognitæ) propoſui tum ſequentia phænomena, tum eorundem cauſas.

Wir haben etliche helle, faſt ætheriſche, olea, die wegen ihrer groſſen Zartheit aufs ſubtileſte flüſſig ſind, helle wie der Himmel, flüchtig wie die Lufft, im Waſſer ſehr unbeſtändig und auf das ſchnelleſte

ver-

A PAGE FROM STAHL'S "FUNDAMENTA CHEMIÆ"

The book begins in Latin and ends in German. The page selected is from the middle portion.

bulky, it attracts water and then gives an acid reaction, hence phosphorus is a compound of phlogiston and phosphoric acid. Sulphur burns completely and we might conclude that sulphur was pure phlogiston were it not for the choking fumes which are evolved. When these are absorbed in water, acids of sulphur are formed which with phlogiston must have made up the original sulphur. When tin and lead are calcined in the air a voluminous ash is formed. The metals, therefore, consist of these calces plus phlogiston. Iron in rusting undergoes a slow combustion and the metal consists of the rust and phlogiston. If the calces of the metals are heated with carbon (which is rich in phlogiston) the carbon gives its phlogiston to the calx and the metal is obtained, a simple explanation of the smelting process. But we need not stop here. The theory may be equally well applied to reactions in the wet way. If iron be introduced into a solution of blue vitriol, it goes into solution while copper is precipitated. Obviously the iron gives up its phlogiston to produce the copper, and we can even draw some conclusions concerning the quantitative relationships involved, for the iron which disappeared and the copper which is precipitated must represent those quantities which contain the same amount of phlogiston. Some of the later believers in the theory took this step. They also recognized the resemblance of the vital processes to combustion and stated that as our bodies are consumed, the lungs constantly exhale phlogiston!

Faults of the Theory.—These illustrations will at least serve to show that a theory is not necessarily true because it can 'explain' a great number of facts. Here was a theory false to the verge of the ludicrous which yet coördinated most facts familiar to the chemists of the day and enabled them to use their knowledge efficiently for the solution of new problems. The phlogiston theory was, therefore, well fitted for its position as a great working hypothesis, and this gave it universal credit in spite of faults so glaring that it is now hard to see why they were not patent to every thoughtful observer.

The faults are themselves instructive. In the first place no one had ever seen any phlogiston, nor could mention a single one of its properties save that it departed on combustion. It was, therefore, a hypothetical substance devised for a single purpose. This, however, troubled no one. It no more occurred to anyone to go out and look for phlogiston than it occurs to us to attempt the isolation of the 'luminiferous ether.'

Another difficulty was that air is required for combustion. How long this had been recognized we do not know. Certainly it was stated in so many words by Basilius Valentinus (see page 14) and must have been generally understood by anyone who could successfully build a fire. The difficulty was met by the statement that the phlogiston did not simply go away in combustion, it united with the air or some portion of it. If there was no air present the fire went out because the phlogiston had nothing with which to combine.

A more serious difficulty lay in the fact observed by at least Boyle, Mayow and Rey[1] that when metals are burned the calx weighs more than the metal, whereas if burning meant a loss of phlogiston it should weigh less. To us this seems insuperable but at this time it received little attention. Few persons made quantitative experiments and those who did, seem to have seriously confused weight with specific gravity, tacitly assuming that a pound of lead must be heavier than a pound of feathers. Others more logical to whom the fundamental facts were brought home, defended their beloved hypothesis with another still more daring. When fire leaves a substance its *upward* flight shows that it possesses the quality of *levity* or negative weight. Unlike all other substances it is not attracted to the center of the earth but repelled from it. Hence the more phlogiston a substance contains the lighter it is! There is, of course, nothing inherently absurd in the idea of something not amenable to the attraction of gravitation, but that just this hypothetical substance should be the only one to show the property might have set men thinking. Unfortunately there is much in a great hypothesis which tends to prevent thinking.

Hoffmann, Boerhave, Marggraf.—A prominent contemporary of Stahl was Friedrich Hoffmann, (1660–1742), who was for some

[1] Rey is sometimes spoken of as though he came near discovering the facts concerning combustion. His charmingly written and amusing paper on *The Increase in Weight of Tin and Lead on Calcination* (1630) has been reprinted by the Alembic Club and everyone who wants to spend a pleasant half hour should read it. No one who does so, however, will be likely to believe the author capable of making any serious discovery.

JOSEPH BLACK
1728–1799

(Facing page 32)

JOSEPH PRIESTLEY
1733-1804

light and that in this form the two substances have left the flask. That heat contained phlogiston would have been considered self-evident in those days, for proof was found in such experiments as that with saltpeter described on a previous page (36), so it only remained for Scheele to show the presence of fire-air and phlogiston in light. The former he does not attempt, but the latter he proves to his satisfaction in the following elegant and original way. He has a sure criterion for phlogiston in the evolution of red fumes with nitric acid; thus calx of copper (copper oxide) evolves no fumes when treated with the acid, but metallic copper (which contains phlogiston) does so copiously. Scheele accordingly exposes silver chloride to light and in that way obtains traces of metallic silver (he was the first to observe this reaction) and he proves its presence by the addition of nitric acid which now gives fumes, whereas the unilluminated silver salt did not. He reasons that the metal obtained its phlogiston from the light (!) since it was the latter which caused the change. Such was the phlogiston theory!

Priestley.—Joseph Priestley, whose discovery of oxygen is an important mile-stone in chemical history, was born March 13, 1733, in Fieldhead, Yorkshire. Priestley's health was delicate in his early years so that his education was obtained mostly by private instruction and suffered many interruptions. Nevertheless he was fond of books and gradually acquired a considerable knowledge of ancient, modern and even oriental languages. He also had some opportunity to study the natural philosophy of those days. He finally decided to enter the ministry and in 1755 began preaching to a small dissenting congregation at Needham Market. This was the first of a series of such pastorates, but Priestley was never a very successful preacher on account of an impediment in his speech, and he did some teaching as opportunity offered, interesting himself more and more in chemical experiments, especially upon gases, a study for which he was eminently fitted by unusual manipulative skill. This scientific work attracted attention and Priestley became a Fellow of the Royal Society in 1766 and a foreign associate of the French Academy of Sciences in 1772. In the latter year he also obtained a congenial position as librarian and literary companion to Lord Shelburne with whom he traveled on the

Continent and thus gained opportunity for contact with the most eminent scientific men in France, including Lavoisier. In 1780 Priestley was pensioned by Lord Shelburne and again began preaching, this time in Birmingham. When the French Revolution broke out he warmly espoused its principles, and in 1791 his attendance at a dinner held to celebrate the anniversary of the fall of the Bastille so infuriated some of his fellow-citizens that they sacked his house and burned the chapel where he preached. Priestley took this as a hint to resign his pastorate, and though he later undertook another charge he found himself so unpopular that in 1794 he decided to abandon England for America. He settled in Northumberland, Pennsylvania, and died there February 6, 1804.

Priestley wrote much on theology and other topics, but his best title to fame is the work recorded in his *Experiments and Observations on Different Kinds of Air*, published between 1774 and 1786. Priestley constantly ascribes his discoveries to chance, and some critics have taken him at his word and condemned his work as planless and haphazard. Impulsive bunglers, however, do not make such discoveries, and we shall come nearer to the truth if we ascribe to Priestley an innocent literary affectation of modesty akin to that found in Montaigne who, for that matter, might well have written the following:

> "I do not think it at all degrading to the business of experimental philosophy, to compare it, as I often do, to the diversion of *hunting*, where it sometimes happens that those who have beat the ground the most, and are consequently the best acquainted with it weary themselves without starting any game; when it may fall in the way of a mere passenger; so that there is but little reason for boasting in the most successful termination of the chase."

Priestley's most conspicuous improvement in the methods of gas-manipulation of his time was the use of mercury instead of water in the pneumatic trough. This enabled him to isolate numerous gases which had hitherto been missed on account of their solubility in water. Among these were sulphur dioxide, hydrochloric acid and ammonia. The last two he designated as "marine acid air" and "alkaline air" respectively. He mingled them in the hope of obtaining a "neutral air" and so synthesized ammonium chloride. This salt had, of course, been known for many centuries. He passed electric sparks through ammonia gas and noted that hydrogen was formed. The exact composition of ammonia was, however, first settled by Berthollet some time after. Priestley also heated fluorspar with sulphuric acid and obtained silicon fluoride. It is interesting that Scheele had also tried this experiment, but because he passed the gases into water he obtained hydrofluorsilicic acid.

Priestley recognized clearly the analogy between combustion and respiration, and, as early as 1772 before he discovered oxygen, he was able to demonstrate experimentally the most important reciprocal relation between animal and plant life, for he found that air in which a candle had been burned until it went out spontaneously, again became respirable and capable of supporting combustion after plants had grown in it for some time. Work of this kind involved experiments with animals, usually mice, and he writes like this concerning his methods of manipulation:

> "For the purpose of these experiments it is most convenient to catch the mice in small wire traps, out of which it is easy to take them, and holding them by the back of the neck, to pass them through the water into the vessel which contains the air. If I expect that the mouse will live a considerable time, I take care to put into the vessel something on which it may conveniently sit, out of reach of the water. If the air be good, the mouse will soon be perfectly at its ease, having suffered nothing by its passing through the water. If the air be supposed to be noxious, it will be proper (if the operator be desirous of preserving the mice for farther use) to keep hold of their tails, that they may be withdrawn as soon as they begin to show signs of uneasiness; but if the air be thoroughly noxious, and the mouse happens to get a full inspiration, it will be impossible to do this before it will be absolutely irrecoverable. . . . Two or three of them will live very peaceably together in the same vessel; though I had one instance of a mouse tearing another almost in pieces, and when there was plenty of provisions for both of them."

The Discovery of Oxygen.—In 1774 Priestley was heating all the substances he could find by means of a large burning lens, and collecting any gases evolved over mercury in the hope of obtaining new gases and observing their properties. It was in this way that he came at last to prepare oxygen. He writes:

"With this apparatus, after a variety of other experiments an account of which will be found in its proper place, on the 1st of August, 1774, I endeavored to extract air from *mercurius calcinatus per se* [mercuric oxide] and I presently found that, by means of this lens, air was expelled from it very readily. Having got about three or four times as much as the bulk of my materials, I admitted water to it, and found that it was not imbibed by it. But what surprised me more than I can well express was that a candle burned in this air with a remarkably vigorous flame, very much like that enlarged flame with which a candle burns in nitrous air, exposed to iron or liver of sulphur; but as I had got nothing like this remarkable appearance from any kind of air besides this particular modification of nitrous air, and I knew no nitrous acid was used in the preparation of *mercurius calcinatus*, I was utterly at a loss how to account for it."

It was some time before Priestley realized that the gas which he had thus isolated was the very component of the atmosphere which supports life and combustion; but Mayow had observed long before that the latter gave red fumes with "nitrous air" (nitric oxide), and Priestley made use of the fact that these fumes (nitrogen peroxide) are soluble in water to make a rough analysis of the air. The new gas, of course, showed the reaction strongly. Concerning its physiological action he writes as follows:

"From the greater strength and vivacity of the flame of a candle in this pure air, it may be conjectured that it might be peculiarly salutary to the lungs in certain morbid cases, when the common air would not be sufficient to carry off the phlogistic putrid effluvium fast enough (see page 29). But, perhaps, we may also infer from these experiments, that though pure dephlogisticated air might be very useful as a *medicine*, it might not be so proper for us in the usual healthy state of the body: for as a candle burns out much faster in dephlogisticated than in common air, so we might, as may be said, *live out too fast*, and the animal powers be too soon exhausted in this pure kind of air. A moralist, at least, may say that the air which nature has provided for us is as good as we deserve.

"My reader will not wonder that, after having ascertained the superior goodness of dephlogisticated air by mice living in it, and the other tests above mentioned, I should have the curiosity to taste it myself. I have gratified that curiosity by breathing it, drawing it through a glass siphon, and, by this means, I reduced a large jar full of it to the standard of common air. The feeling of it to my lungs was not sensibly different from that of common air; but I fancied that my breath felt peculiarly

PRIESTLEY'S LABORATORY

light and easy for some time afterward. Who can tell but that, in time, this pure air may become a fashionable article in luxury? Hitherto only two mice and myself have had the privilege of breathing it."

We have seen that Priestley called the gas he had discovered dephlogisticated air, his idea being that this was the component of the atmosphere with which the phlogiston united when it emerged from a burning substance. He called nitrogen "phlogisticated air," and this nomenclature would seem to imply that he considered it a product of such union. If so nitrogen should sometimes appear as a product of combustion, but this contradiction was overlooked, like every fact which told against the phlogiston theory. Like Scheele, Priestley missed entirely the real significance of his discovery. Both were so sure that something was always *given off* in combustion that they had lost the power to believe that the burning body *united with* one of the gases of the atmosphere even when they saw the latter disappear before their eyes. Such blindness was really less pardonable in Priestley than in any of the others, for he not only could not draw the correct conclusion from his own experiments, but all the brilliant work of Lavoisier a little later failed utterly to convince him, and he defended the theory of phlogiston to the last. As late as 1800 he wrote to a friend:

"I have well considered all that my opponents have advanced, and feel perfectly confident of the ground I stand upon. . . . Though nearly alone I am under no apprehension of defeat."

Literature

BLACK'S *Experiments upon Magnesia, Quicklime, and some other Alkaline Substances* has been reprinted by the Alembic Club. His lectures were published with a biographical preface by JOHN ROBISON in 1803. SIR WILLIAM RAMSAY devotes a chapter to Black in his *Essays Biographical and Chemical,* London, 1908.

A *Life of Cavendish* by GEORGE WILSON was published by the Cavendish Society, London, in 1851. See also THORPE'S *Essays.* His experiments on air may be found in the *Alembic Club Reprints* No. 3.

An English edition of SCHEELE'S *Chemical Essays* was issued by THOMAS BEDDOES in 1786, and one of his *Air and Fire* by J. R. FOSTER in 1785. The latter is also reprinted in OSTWALD'S *Klassiker* No. 58. SCHEELE'S account of his experiments bearing on the discovery of oxygen are to be found in *Alembic Club Reprint* No. 8. His laboratory notebooks and other memoranda were collected by NORDENSKIÖLD in 1892 and published under the title *Nachgelassene Briefe und Aufzeichnungen von Karl Wilhelm Scheele.*

PRIESTLEY's *Experiments and Observations on Different Kinds of Air* is still to be found in some libraries, and is most interesting reading. The portion of it bearing directly on the discovery of oxygen can be found in *Alembic Club Reprint No. 7.* There is an attractive life of Priestley by Thorpe, London, 1906, who also devotes a chapter to him in his *Essays.*

ɔnstitute a table of affinity for the given base. A similar might, of course, be constructed for an acid or indeed any substance and would take a form like the following:

Fixed alkali	Vitriolic acid
Vitriolic acid	Fixed alkali
Acid of nitre	Volatile alkali
Marine acid	Absorbent earth
Acid of vinegar	Iron
Sulphur	Copper
	Silver

:oy's assumption was an attractive one for it seems easy ieve that if, in the case of a given compound, we knew all ıbstances which had affinity for it, and the relative degree ,t affinity, the chemistry of the substance would be thereby etely determined. Today, however, we know that temıre, pressure, solubility and the nature of the medium have ıch to do with the course of chemical reactions that such could never do justice to more than a portion of the truth. 'theless, they served their day as a convenient and compact :or collating chemical facts.

ıelle.—Another eminent French teacher of chemistry was ,ume François Rouelle (1703–1770), who should be mentioned were it only for the fact that he numbered Lavoisier among ıpils. He was, however, himself a scientist of no mean r and fixed more clearly than any had done before him the ɔf a salt as an addition product of acid and base (not acid ıetal), and he distinguished between neutral, acid and basic something which often mystifies beginners even in our ımes.

Literature

MAS BIRCH published the complete works of Boyle in six imposing ɜs. The second edition appeared in 1772. There is an attractive t of his life and work in THORPE'S *Essays in Historical Chemistry*, ;ion, London, 1902.

'ow's studies of the *Spiritus Nitro-Aerius* have been published by .LD in his *Klassiker der Exakten Wissenschaften*, No. 126.

ılso the fascinating paper by JEAN REY, *Alembic Club Reprints* No. 11.

years professor at Halle and a vigorous and voluminous writer. His experimental investigations were largely concerned with the examination of mineral waters, and in testing for their various constituents he did much toward perfecting the analytical methods of his time. Another contemporary was Hermann Boerhave (1668–1738), an eminent professor at the University of Leiden, who acquired fame as a teacher both in medicine and chemistry. His great work entitled *Elementa Chemiæ,* was published in 1724 and long ranked as chief authority upon the subject. Among the most prominent of the German phlogistians was Andreas Sigismund Marggraf (1709–1782), who was for many years director of the laboratory of the Academy of Sciences in Berlin. Marggraf made some important discoveries, among which may be mentioned the distinction of magnesia from alumina, and the use of the flame coloration for distinguishing soda from potash. He recognized that gypsum, barytes and potassium sulphate are all derivatives of sulphuric acid, and observed that phosphorus gains in weight when burned. This, however, did not shake his allegiance to the phlogiston theory. Marggraf made considerable use of the microscope, and by its means detected the presence of sugar in the beet, an observation destined to bear fruit industrially.

Geoffroy and His Tables of Affinity.—Among the French chemists of this period Étienne François Geoffroy (1672–1731) is particularly worthy of mention. He was a prominent lecturer at the *Jardin du Roi* and the *Collège de France* and did a signal service by his *Tables of Affinity* which were presented to the Academy of Science between 1718 and 1720. The fundamental idea underlying these tables was the following: If we consider a given base, say caustic potash, we recognize that it reacts vigorously with a variety of substances, especially with acids, and it is natural to inquire which of these has the greatest affinity for the base. Geoffroy tried to answer this question by the plausible assumption that when an acid has combined with a base and the product is brought into contact with a second acid, the latter, if it has a greater affinity for the base than the first, will expel it. For every base, therefore, it should be possible to prepare a list of acids in such order that any one substance will expel from combination all those which succeed it. This

ANTOINE LAURENT LAVOISIER
1743-1794

(Facing page 46)

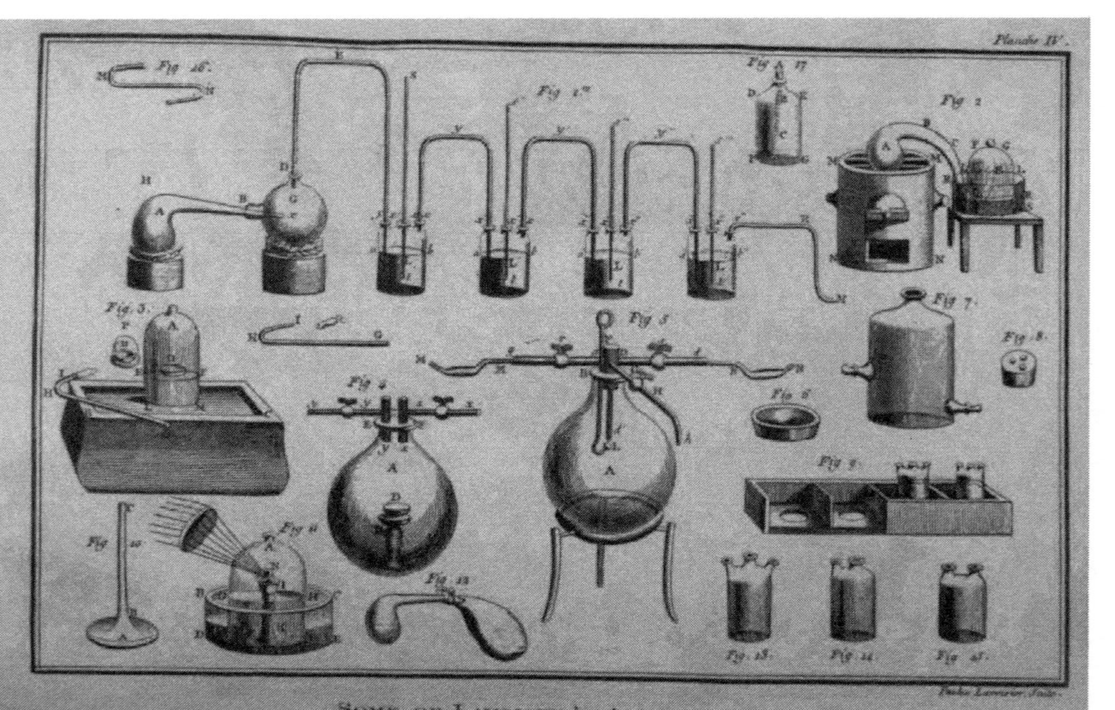

SOME OF LAVOISIER'S APPARATUS

CHAPTER VI

LAVOISIER

Antoine Laurent Lavoisier was born in Paris, August 26, 1743, and began his studies at the *Collège Mazarin* where he came in contact with some distinguished teachers, notably Rouelle in chemistry. As early as 1766 a gold medal was awarded him by the Academy of Sciences in recognition of a paper dealing with the problem of lighting a large town. In 1768 he was made an *adjoint chimiste* to the Academy and began to present frequent reports upon the greatest variety of topics. In 1769 he became associated in a subordinate capacity with the Farmers General of the revenue and soon after was made a member of the board. In 1775 he was appointed *régisseur des poudres* and made valuable suggestions for the improvement of the product. The foregoing, however, represent only a portion of his public activities. We find him serving with tireless energy on all sorts of boards and commissions both national and municipal, solving with equal skill troublesome problems of administration like taxation, banking, coinage, public charity and scientific agriculture. With the outbreak of the Revolution these activities at first increased, but the board of Farmers General gradually became objects of suspicion and rumors of peculation were circulated concerning them. In 1791 they were suppressed and Lavoisier's administration of the *régie des poudres* attacked by Marat to whom Lavoisier had been so unfortunate as to give personal offence. In November, 1793, the Farmers General were put under arrest, and in May of the following year they were sent for trial before the Revolutionary Tribunal, one of the principal charges against Lavoisier being that he had put water in the soldiers' tobacco. The result of the trial was the usual foregone conclusion and Lavoisier was condemned to death by the guillotine within twenty-four hours, the judge cynically remarking: "*La République n'a pas besoin de savants.*" The sentence was carried out on

May 8, 1794, and the feeling of posterity concerning the wrong thus done the world found fit expression at the time in the bitter words of Lagrange: "It took them but a moment to cut off that head, though a hundred years, perhaps, will be required to produce another like it."

Temperament and Method of Investigation.—Lavoisier brought to the study of chemistry the equipment most needed at this time—the habits and mental attitude of the trained physicist. We shall often have occasion to see that chemistry has gained enormously by the influence of those whose point of view has been preëminently physical, men who do not care to prepare new compounds or discover new reactions, but who prefer to weigh and measure, and in this way gain insight into the mechanism of chemical changes already familiar on the qualitative side. Lavoisier is one of the most conspicuous examples of this type of mind while Scheele admirably represents the opposite extreme, and this makes it especially interesting to see how the two men once solved the same problem, each in a manner characteristically his own.

The alchemists had observed that when water is boiled for some time in a glass or earthen vessel it can no longer be distilled without leaving a solid residue, and if the boiling be really long-continued a sediment appears. They interpreted this as a transformation of *water* into *earth* under the influence of *fire*. The fundamental fact had been confirmed by Boyle, Boerhave, and many others, but the explanation seemed essentially improbable to Lavoisier and about 1770 he demonstrated its falsity by the following experiment. He boiled a *weighed* quantity of water in a *weighed* 'pelican' for a hundred days. The pelican was a closed vessel having a long neck bent back upon itself which served the same purpose as the modern reflux condenser. At the end of the long boiling the total weight of the flask and its contents had not changed showing that nothing had been lost or received from the fire. The weight of the flask, however, had diminished, while the sum of the weights of water and sediment was greater than that of the original water by a practically equivalent amount, showing that the solid material had come from the glass and not from the water. Scheele proceeded differently. He also boiled water a long time in glass, but he weighed nothing. Instead he tested the sediment qualitatively, and finding that it contained potash, silica and lime, which he knew to be constituents of glass, he came to the same conclusion as Lavoisier. Who can say that one method was superior to the other? At the present time, of course, we use one method to supplement the other as occasion demands, but when it comes to original work it still holds true that the great advances in one line are rarely made by the same men who make great advances in the other. The two types of investigators remain distinct.

Work on Combustion.—Lavoisier's work upon the nature of combustion followed naturally from that just described. Here also he burns weighed quantities of material and carefully weighs the products, and from his results he is able to draw far-reaching conclusions. On November 1, 1772, he handed to the secretary of the Academy a sealed note which read as follows:

"About eight days ago I discovered that sulphur in burning, far from losing weight, rather gains it; that is to say that from a pound of sulphur may be obtained more than a pound of vitriolic acid, allowance being made for the moisture of the air. It is the same in the case of phosphorus. The gain in weight comes from the prodigious quantity of air which is fixed during the combustion and combines with the vapors.

"This discovery, which I have established by experiments which I consider decisive, has made me believe that what is observed in the combustion of sulphur and phosphorus may equally well take place in the case of all those bodies which gain weight on combustion or calcination. I am persuaded that the gain in weight of the metallic calces is due to the same causes. Experiment has completely confirmed my conjectures. I have reduced litharge in closed vessels, employing the apparatus of Hales, and I observed that a considerable quantity of air is evolved just at the moment when the litharge changes to metal, and that this air occupies a volume a thousand times greater than the quantity of litharge employed.

"Since this discovery seemed to be one of the most interesting which has been made since the time of Stahl I have felt it my duty to place this communication in the hands of the secretary of the Academy, to remain a secret until I can publish my experiments."

Theory of Combustion.—It is clear from the above that Lavoisier now fully realized the important consequences which must follow if combustion really proved to be a union of the burning substance with the air or some part thereof. One experiment now follows another rapidly and in the fall of 1774 Lavoisier publishes his work on the calcination of tin. He takes a weighed portion of the metal, seals it in a weighed flask large enough to contain considerable air and heats the whole until the metal appears well calcined. When cold he finds that the system has neither gained nor lost in weight, and he then breaks the seal. A partial vacuum is revealed by air rushing in, and he now finds the weight of the flask increased by exactly the same amount as the tin itself has gained, showing that the calcination consisted in a transfer of gas from the air to the tin. He further finds that when sufficient tin is used an air is left which will calcine no more tin, and from this he concludes that only a part of the air reacts in combustion. Still later in the same year, after the first experiments of Priestley, he heats mercury in a limited volume of air until a considerable quantity of the red oxide has been formed, and notes the decrease in the volume of air, which finally reaches a maximum. He next like Priestley heats the mercuric oxide alone and fixes with precision the properties of the gas evolved and the vigor with which it supports combustion.

In 1777 he sums up his theory of combustion in the following four propositions:

1. In every combustion heat and light are evolved.
2. Bodies burn only in *air éminemment pur* (this was Lavoisier's first name for oxygen).
3. The latter is used up by the combustion, and the gain in weight of the substance burned is equal to the loss of weight shown by the air.
4. By the process of combustion the combustible substance is usually changed to an acid; the metals, however, are calcined.

The Chemical Revolution.—In spite of the clearness of Lavoisier's presentation, and the skilful experiments by which his conclusions were supported, his views at first made little impression until another series of experiments carried on in 1782 and 1783 led to a conclusion of a similar kind. Lavoisier turned his attention to the composition of water, and though his experiments followed those of Cavendish, yet he was able to interpret them as Cavendish could not. Original was his quantitative decomposition of steam by passing it over hot iron. He also burned oxygen and hydrogen, forming water synthetically and measuring all quantities involved. Still later he experimented in the same way with carbonic acid, found that it was formed by the union of carbon and oxygen and fixed its composition by weight. He also confirmed what Black had done on the causticization of lime. The cumulative effect of these researches led to the sudden conversion of most French chemists about 1785, and the new ideas were firmly fixed by the publication of Lavoisier's great text-book *La Traité Élémentaire de la Chimie* in 1789. The change was well called the Chemical Revolution, for it inverted completely the chemical point of view. The mysterious hypothetical substance, phlogiston, which did not obey the law of gravitation, and changed its properties arbitrarily as theoretical considerations dictated, was banished from the science and the law of conservation of mass vindicated once for all. This made possible quantitative analysis and the chemical equation. The latter is a convenience in form of expression which we owe to Lavoisier, and of whose assistance in stating and solving chemical problems he made immediate use. He writes:[1]

"If I distil an unknown salt with vitriolic acid and find nitric acid in the receiver and vitriolated tartar in the residue, I conclude that the original salt was nitre, and I reach this conclusion by mentally writing the following equation based upon the supposition that the total weight is the same before and after the operation.

"If x is the acid of the unknown salt, and y is the unknown base I write:

$x + y +$ vitriolic acid = nitric acid + vitriolated tartar = nitric acid + vitriolic acid + fixed alkali.

"Hence I conclude: $x =$ nitric acid, $y =$ fixed alkali, and the original salt was nitre."

Nothing like this had ever appeared before in chemical literature and what a fog of mystery and superstition it removed!

Lavoisier was a pioneer in the analysis of organic substances, which he burned in air or oxygen and determined the weight of carbonic acid formed. He had some trouble with his absorption

[1] The paragraph included in the quotation marks is a much abridged paraphrase.

apparatus which was rather crude but many of his analyses were surprisingly good. He also, like Priestley, appreciated the analogy between respiration and combustion and made quantitative experiments to determine the rate of the oxidation in the animal body. This he assumed to take place in the lungs. We may consider this work as the beginning of physiological chemistry.

The New Nomenclature.—In 1782, Guyton de Morveau (1737-1816) published in the *Journal de Physique* a paper suggesting reforms in chemical nomenclature. The work of Lavoisier had made the time ripe for such a movement and a commission in which Guyton de Morveau was associated with Lavoisier, Fourcroy and Berthollet was appointed to consider the whole matter. In 1787 the results of their deliberations were published under the title, *Méthode d'une Nomenclature Chimique.* This work did away with many of the fanciful and arbitrary names previously in use, and substituted such as were based on chemical composition. Speaking broadly it is hardly too much to say that it laid the foundation upon which our modern international nomenclature now stands.

Chemical Knowledge Before the Revolution.—Books written from the phlogistian point of view are now so difficult to read that we are inclined to underestimate the chemical knowledge of that period, so that a glance at the tables of Lavoisier's *Traité Élémentaire* which contrast the old names with the new may furnish a wholesome corrective. It is true, of course, that up to this time the subject of chemical composition had been in the utmost confusion, for if a metal was to be considered as a compound of its calx with phlogiston, and the compound weighed less than the sum of its component parts, then such words as element and compound had no significance, and as a matter of fact the terms do not seem to have been much contrasted. Nevertheless, certain things were clear. The differences which distinguish acids, bases and salts could not be overlooked. Salts were regarded as addition products of acid and base, and some of the newest and less familiar were already named as such. Acid and basic salts were already distinguished from neutral ones by the excess of one component, and emphasis had been laid upon the fact that it was the base and not the metal which

TABLEAU des Combinaisons binaires de l'oxygène avec les substances métalliques & non métalliques oxidables & acidifiables.

		Premier degré d'oxygénation.		*Second degré d'oxygénation.*		*Troisième degré d'oxygénation.*		*Quatrième degré d'oxygénation.*	
		Noms nouveaux.	*Noms anciens.*	*Noms nouveaux.*	*Noms anciens.*	*Noms nouveaux.*	*Noms anciens.*	*Noms nouveaux.*	*Noms anciens.*
Combinaisons de l'oxygène avec les substances simples non métalliques, telles que :	Le calorique	Le gaz oxygène	Air vital ou déphlogistiqué						
	L'hydrogène	On ne connoît qu'un degré de combinaison de l'oxygène & de l'hydrogène, & cette combinaison forme de l'eau							
	L'azote	Oxide nitreux ou base du gaz nitreux	Gaz nitreux	Acide nitreux	Acide nitreux fumant	Acide nitrique	Acide nitreux non fumant	Acide nitrique oxygéné	Inconnu.
	Le carbone	Oxide de carbone	Inconnu	Acide carboneux	Inconnu	Acide carbonique	Air fixe	Acide carbonique oxygéné	Inconnu.
	Le soufre	Oxide de soufre	Soufre mou	Acide sulfureux	Acide sulfureux	Acide sulfurique	Acide vitriolique	Acide sulfurique oxygéné	Inconnu.
	Le phosphore	Oxide de phosphore	Résidu de la combustion du phosphore	Acide phosphoreux	Acide volatil du phosphore	Acide phosphorique	Acide phosphorique	Acide phosphorique oxygéné	Inconnu.
	Le radical muriatique	Oxide muriatique	Inconnu	Acide muriateux	Inconnu	Acide muriatique	Acide marin	Acide muriatique oxygéné	Acide marin déphlogistiqué.
	Le radical fluorique	Oxide fluorique	Inconnu	Acide fluoreux	Inconnu	Acide fluorique	Inconnu des anciens	…	
	Le radical boracique	Oxide boracique	Inconnu	Acide boraceux	Inconnu	Acide boracique	Sel sédatif de Homberg	…	
Combinaisons de l'oxygène avec les substances simples métalliques, telles que :	L'antimoine	Oxide gris d'antimoine	Chaux grise d'antimoine	Oxide blanc d'antimoine	Chaux blanche d'antimoine. Antimoine diaphorétique	Acide antimonique	…	…	
	L'argent	Oxide d'argent	Chaux d'argent	…	…	Acide argentique	…	…	
	L'arsenic	Oxide gris d'arsenic	Chaux grise d'arsenic	Oxide blanc d'arsenic	Chaux blanche d'arsenic	Acide arsenique	Acide arsenical	Acide arsenique oxygéné	Inconnu.
	Le bismuth	Oxide gris de bismuth	Chaux grise de bismuth	Oxide blanc de bismuth	Chaux blanche de bismuth	Acide bismuthique	…	…	
	Le cobalt	Oxide gris de cobalt	Chaux grise de cobalt	…	…	Acide cobaltique	…	…	
	Le cuivre	Oxide rouge brun de cuivre	Chaux rouge brune de cuivre	Oxide vert & bleu de cuivre	Chaux verte & bleue de cuivre	Acide cuprique	…	…	
	L'étain	Oxide gris d'étain	Chaux grise d'étain	Oxide blanc d'étain	Chaux blanche d'étain ou potée d'étain	Acide stannique	…	…	
	Le fer	Oxide noir de fer	Ethiops martial	Oxide jaune & rouge de fer	Ocre & rouille	Acide ferrique	…	…	
	La manganèse	Oxide noir de manganèse	Chaux noire de manganèse	Oxide blanc de manganèse	Chaux blanche de manganèse	Acide manganique	…	…	
	Le mercure	Oxide noir de mercure	Ethiops minéral	Oxide jaune & rouge de mercure	Turbith minéral, précipité rouge, précipité per se	Acide mercurique	…	…	
	Le molybdène	Oxyde de molybdène	Chaux de molybdène	…	…	Acide molybdique	Acide de la molybdène	Acide molybdique oxygéné	Inconnu.
	Le nickel	Oxide de nickel	Chaux de nickel	…	…	Acide nickelique	…	…	
	L'or	Oxide jaune d'or	Chaux jaune d'or	Oxide rouge d'or	Chaux rouge d'or. Précipité pourpre de Cassius	Acide aurique	…	…	
	La platine	Oxide jaune de platine	Chaux jaune de platine	…	…	Acide platinique	…	…	
	Le plomb	Oxide gris de plomb	Chaux grise de plomb	Oxide jaune & rouge de plomb	Massicot & minium	Oxide plombique	…	…	
	Le tungstène	Oxide de tungstène	Chaux de tungstène	…	…	Acide tungstique	Acide de la tungstène	Acide tungstique oxygéné	Inconnu.
	Le zinc	Oxide gris de zinc	Chaux grise de zinc	Oxide blanc de zinc	Chaux blanche de zinc, Pompholix	Acide zincique	…	…	

TABLE FROM LAVOISIER'S *Traité Elementaire* ILLUSTRATING THE NEW NOMENCLATURE

DES SUBSTANCES SIMPLES.

TABLEAU DES SUBSTANCES SIMPLES.

	Noms nouveaux.	*Noms anciens correspondans.*
Substances simples qui appartiennent aux trois règnes, & qu'on peut regarder comme les élémens des corps.	Lumière	Lumière.
	Calorique	Chaleur.
		Principe de la chaleur.
		Fluide igné.
		Feu.
		Matière du feu & de la chaleur.
	Oxygène	Air déphlogistiqué.
		Air empiréal.
		Air vital.
		Base de l'air vital.
	Azote	Gaz phlogistiqué.
		Mofète.
		Base de la mofète.
	Hydrogène	Gaz inflammable.
		Base du gaz inflammable.
Substances simples non métalliques oxidables & acidifiables.	Soufre	Soufre.
	Phosphore	Phosphore.
	Carbone	Charbon pur.
	Radical muriatique	Inconnu.
	Radical fluorique	Inconnu.
	Radical boracique	Inconnu.
Substances simples métalliques oxidables & acidifiables.	Antimoine	Antimoine.
	Argent	Argent.
	Arsenic	Arsenic.
	Bismuth	Bismuth.
	Cobalt	Cobalt.
	Cuivre	Cuivre.
	Etain	Etain.
	Fer	Fer.
	Manganèse	Manganèse.
	Mercure	Mercure.
	Molybdène	Molybdène.
	Nickel	Nickel.
	Or	Or.
	Platine	Platine.
	Plomb	Plomb.
	Tungstène	Tungstène.
	Zinc	Zinc.
Substances simples salifiables terreuses.	Chaux	Terre calcaire, chaux.
	Magnésie	Magnésie, base du sel d'epsom.
	Baryte	Barote, terre pesante.
	Alumine	Argile, terre de l'alun, base de l'alun.
	Silice	Terre siliceuse, terre vitrifiable.

LAVOISIER'S TABLE OF THE ELEMENTS

ned the salt by direct addition. When a metal did dissolve acid to form a salt it must first give up phlogiston (evolve lrogen) as zinc did in muriatic acid, or else the acid must lf be 'phlogisticated' (reduced) as when copper reacts with phuric or nitric acids. What we should now call differences state of oxidation were also recognized, and nitrates disguished from nitrites, manganates from manganous salts and like.

mong organic compounds no rational system was yet possible. ll-known substances like sugar and alcohol had of course ir present names, but these were then devoid of general mical significance; the majority of such compounds had names gestive of the source from which they had been derived like of turpentine' or 'sweet spirits of nitre.' How far this nt is illustrated by the old tradition (which the author has er been able to verify) that in one book of this period cow's ter is found classified with 'butter of antimony' $SbCl_3$. vertheless a good many substances had received extensive alitative study and the field had recently been much enriched the work of Scheele upon the vegetable acids.

avoisier's Table of Elements.—Lavoisier's *Traité Élémenre* was originally begun in order to expound and promulgate new nomenclature, and one of the most interesting things in book is the table of the elements, for it admirably reveals author's point of view. Following Boyle, he defines these substances which cannot be further decomposed, and divides em into four groups. The first comprises the elementary ses, oxygen, hydrogen and nitrogen along with heat and ht. The second contains those elements like sulphur and osphorus which on oxidation yield acids—elements which we w classify as metalloids. In the third group are the metals, ile the fourth is made up of the 'earths'; lime, magnesia, ryta, alumina and silica. These of course could not be classid otherwise since they had not as yet been decomposed. The ne might have been said of the alkalies; soda and potash, but voisier is so certain that these are oxides of "radicals" soon be discovered that he explicitly refuses to include them.

The modern reader is apt to feel surprise in finding heat and ht in the list. We have to remember, however, that our modern conception of energy and its transformations was then unknown, and that men like Scheele were still explaining phenomena by considering heat and light as corporeal substances. Lavoisier did not do this. He recognized that they did not possess weight, but he also realized how constantly their appearance is associated with chemical change, and he regarded heat as a kind of atmosphere which surrounds the ultimate particles of all bodies causing repulsion and hence expansion. He was also familiar with the disappearance of heat when a substance passes from the solid to the liquid, or from the liquid to the gaseous condition, and he was inclined to interpret this chemically as a combination of heat with the substance. Similarly in regard to light, we find the statement that silver salts probably absorb it, and that light also combines with growing plants, since they die or change their properties when deprived of it.

We have already seen how the old system had magnified the importance of phlogiston and made it the center of all chemical explanations. When Lavoisier substituted reduction for phlogistication and oxidation for dephlogistication it was only natural that the newly discovered element oxygen should usurp the position of exaggerated importance from which phlogiston had just been displaced. This is exactly what happened. Every element found its position in the system of Lavoisier according to its relation toward oxygen. Metals had hitherto been compounds of bases with phlogiston. They now became the elements which united with oxygen to form bases. The metalloids on the other hand became the elements which united with oxygen to form acids, indeed the very name *oxygen* or "acid former" was intended to express this fact.

The Oxygen Theory of Acids.—It seemed to follow naturally that all acids must contain oxygen, and Lavoisier did not hesitate to draw this conclusion. Boric acid was therefore to him the oxide of a "radical" as yet unknown, and for this radical he finds a place in the table of the elements. In so doing he was justified by subsequent events, but the fact that he made the same assumption in the case of hydrochloric and hydrofluoric acids was destined later to prove most unfortunate. One minor difficulty arose at the start. If there is no phlogiston, and oxygen is the essential component of acids, why do they evolve *hydrogen* with metals? An 'explanation' was soon found, however, to this effect: Water is simultaneously decomposed; the hydrogen is evolved as such, while the oxygen unites with the metal to form a base, which now can add an acid to form a salt. What a debt all chemical theorists owe to water!

Lavoisier could not, of course, reduce organic chemistry to a system, but he did apply the same principle of classification to the organic acids. He knew how frequently these could be prepared by processes of oxidation, as acetic acid for example is prepared from alcohol, or oxalic from cane sugar. He therefore regarded these acids also as oxygen compounds of "*compound radicals*" which latter he thought of as being usually composed of carbon and hydrogen, but which might sometimes contain also phosphorus, nitrogen or sulphur. He even suspected the presence of nitrogen in some acids like acetic and tartaric which we now know do not contain it. It will repay us to note carefully the sense in which Lavoisier uses the word *radical*. To him it signifies the element or group of elements which enters into combination with oxygen. It is a word which has changed its meaning frequently since his day.

Literature

Grimaux, *Lavoisier*, 1743–1794, *d'après sa Correspondance ses Manuscrits*, etc., Paris, 1888, is the principal biography, but contains comparatively little on the chemical side. Lavoisier's complete works were published under the auspices of the Minister of Public Instruction in 6 volumes between 1864–1893. There are numerous appreciations. See particularly Berthelot, *La Révolution Chimique*, 1890, and the *Essay* by Thorpe which presents a less flattering view.

CHAPTER VII

THE LAW OF DEFINITE PROPORTIONS

The work of Lavoisier had supplied a theoretical basis for quantitative analysis and he himself did some pioneer work in this branch of the science.

Improvements in technique were soon after introduced by such men as Vauquelin (1763–1829) in France and Klaproth (1743–1817) in Germany. Meantime no one questioned the universal tacit assumption that the percentage composition of a chemical compound is always the same, regardless of its source. We now call this generalization the *law of definite proportions*, and recognize it as a corner-stone of the science. At the end of the eighteenth century, however, it was not adequately supported by experiment and good was done through the attack upon it made by Berthollet.

Berthollet.—Claude Louis Berthollet was born in Talloire, Savoy in 1748. He came to Paris at the age of twenty-four and began the practice of medicine, in which he was very successful. In 1780 he was made a member of the Academy, and about the same time gave up his practice in order to devote himself to work for the government. We have already seen how he coöperated with Lavoisier in the reform of chemical nomenclature. Later he was a professor at the *École normale*. Berthollet became associated with Napoleon at the time that works of art were being removed from Italy to France; he gave him instruction in chemistry, and later accompanied him to Egypt. Berthollet received many honors during the Consulate and Empire, and was made a peer by the Bourbons after the restoration. He died at Arceuil in 1822. His chief service to industry was the introduction of chlorine as a bleaching agent—a discovery which he declined to patent.

Statique Chimique.—It was while in Egypt in 1799 that he presented to the Institute in Cairo an article upon *The Laws*

Berthollet Visits Lavoisier at the Laboratory of the Sorbonne

(Facing page 60)

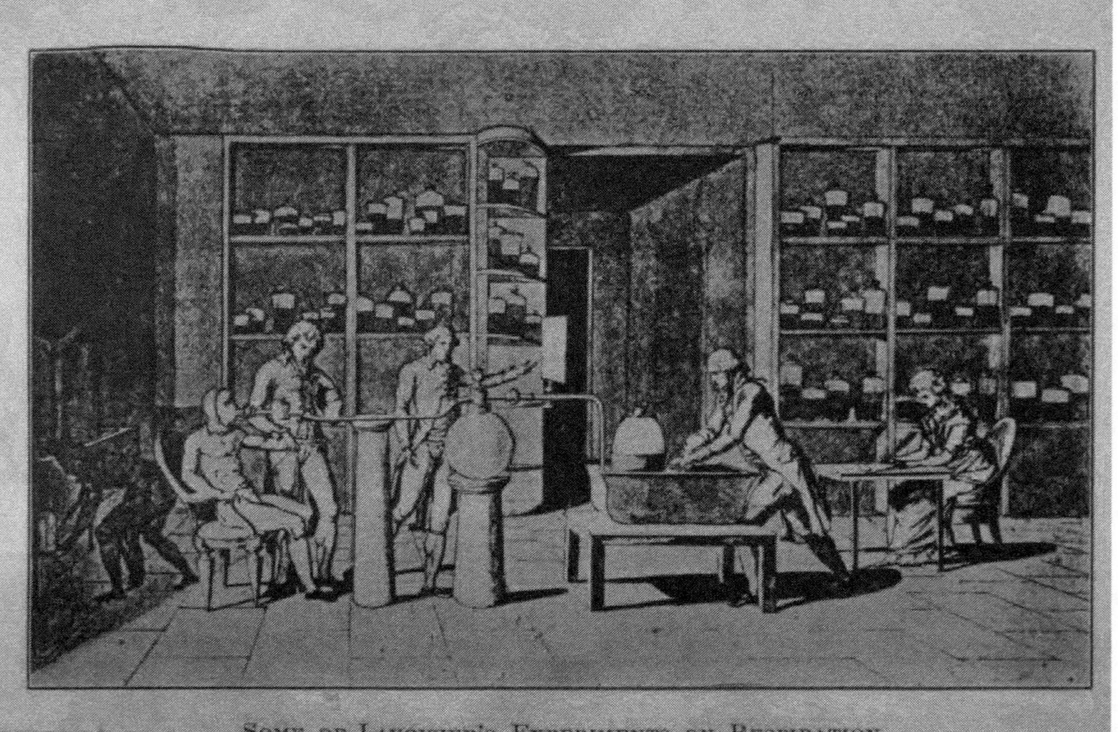

Some of Lavoisier's Experiments on Respiration

of Chemical Affinity. This set forth the fundamental ideas which he later included in his famous book, *Essai de Statique Chimique.* Berthollet began with an attack upon the tables of affinity (page 31). We have already seen how these were originated by Geoffroy and some consideration has been given to their fundamental faults. In spite of this the tables had remained in use as a convenient form for registering chemical information. Lavoisier employed them freely in his *Traité Élémentaire,* and Bergmann in Sweden had enlarged them so that they would be applicable within a wider range of experimental conditions. Berthollet condemned them as wrong in principle. If there were any truth in them then affinity should be something absolute, and one acid should expel another completely from its salts. Berthollet could show that this was rarely the case. Instead, the *quantities* involved play an extremely important rôle, and by adding a sufficient quantity of one reagent the apparent course of a reaction can frequently be reversed. The familiar decomposition of steam by iron and the reduction of iron oxide by hydrogen is a case in point, but Berthollet could cite many other examples. He pointed out that when two salts are mixed the acid and basic components are in an equilibrium in the solution, and the reaction which may take place depends in part upon the affinities, in part upon the masses present, and in part upon other secondary considerations which are nevertheless decisive in certain cases, viz., solubility or volatility of one of the reacting combinations. Thus if sulphuric acid expels carbonic from its salts, it is not because of its greater affinity for the base, but because the carbonic acid is volatile and cannot remain to compete. If sulphuric acid decomposes barium chloride it is again not so much a matter of affinity as of the insolubility of barium sulphate. We see how close these ideas come to our modern notions of equilibrium and reversible reaction. Indeed if Berthollet had been a little clearer as to the importance of concentration as distinguished from mere quantity he would have come very close to the law of mass action. Unfortunately for the success of his own contentions, but doubtless fortunately for the progress of the science, he felt that his observations led him to certain other conclusions not so well justified by the facts. It seemed natural for him to conclude that within certain limits elements ought to unite in all proportions, and he published analyses which went to show that they did so. From this point of view it seemed natural to expect that copper and sulphur, for example, would combine in varying proportions according to which element was present in excess. He recognized, it is true, that some well-characterized compounds seemed to show constant composition, but he attributed this to the insolubility of certain combinations or to their peculiar capacity for crystallization. Finally, he pointed out that no sharp line can be drawn between what we call chemical union and solution, especially as the latter is illustrated in the alloys and glasses. If Berthollet had kept attention focussed upon these latter forms of combination it seems not unlikely that he might have maintained his views longer than he did. As a matter of fact, he chose to meet the issue frankly upon the composition of salts and oxides, where he was open to attack.

Controversy with Proust.—The attack came from Joseph Louis Proust (1755–1826), a talented Frenchman then professor in Madrid. There resulted a controversy in the journals which lasted nearly eight years and was carried on by both sides with the greatest brilliancy and most unfailing courtesy. At the end Proust was left in possession of the field, for wherever Berthollet supposed that he had prepared a series representing all degrees of composition, Proust was able to show that there were really always involved mixtures of *two* substances, each of which when pure had different properties and showed *constant* chemical composition. Proust is therefore to be regarded as the discoverer of the law of definite proportions.[1]

The duke of Wellington is credited with saying: "There is nothing worse than a great victory except a great defeat," and this is often true of scientific controversies. The defeat of Berthollet meant that the good in his contentions was discredited along with the bad, and in consequence little attention was paid to chemical equilibrium and mass action for more than

[1] How near he came to discovering the law of multiple proportions is shown by the fact that he went so far as to calculate the different amounts of oxygen which unite with 100 parts of copper in the two oxides. The correct figures are 12.6 and 25.2 which would have revealed the law, but owing to faulty analyses Proust obtained 18 and 25.

ifty years. This, however, is not an occasion for so much egret as it might seem. The time was not ripe. It was necessary hat other important advances should be made both in chemistry .nd physics before the ideas of Berthollet could really become ruitful. On the other hand, the law of definite proportions vas something the world needed at the moment, and if it had een discredited in 1799 the progress of the science would have een seriously retarded.

Neutralization Phenomena.—Before the controversy between roust and Berthollet began, an interesting study which ap- roached the subject of chemical composition from another point f view was being made in Germany. We have seen what a liscouragement the phlogiston theory had been to all quantita- ive thinking. Nevertheless even the earliest chemists had made ome quantitative observations. From the time of Geber it ad been recognized, for example, that a certain *quantity* f acid was necessary in order to neutralize a given amount f base, and in 1699 Homberg had attempted to determine he quantities of several different acids which neutralized an unce of potash. Glauber is credited with the important tatement that when two neutral salts are mixed, the solution emains neutral whether any chemical reaction is observable or ot. About 1775 both Bergman and Kirwan tried to determine he amounts of different acids which neutralize a given quantity f the same base, and the reverse, but owing to crude experimental nethods could deduce no generalization. Bergman, however, oticed that when copper is immersed in a neutral silver nitrate olution, silver is deposited and copper dissolves, while the solution emains neutral. He concluded that the weights of dissolved opper and precipitated silver are the quantities of those metals which contain the same amount of phlogiston (see page 29). avoisier reversed this, stating that the weights of two metals which dissolve in a given quantity of the same acid, are the veights which unite with the same quantity of oxygen.

Richter.—The credit for combining these observations into system was reserved to Jeremias Benjamin Richter (1762– 807) who at the time this work was done was 'arkanist' at ie royal porcelain factory in Berlin. He published two books, ealing with this subject; one, *Anfangsgründe der Stöchiometrie,* appeared serially between 1792 and 1794, the other, *Ueber die Neueren Gegenstände in der Chemie,* between 1792 and 1802. Richter was particularly impressed by the fact that two neutral salts can react without loss of neutrality. He saw that if sulphate of potash, for example, reacts with nitrate of lime to form nitrate of potash and sulphate of lime, then the preservation of the neutral reaction requires that the *amount* of lime which passes from the nitric to the sulphuric acid must be chemically equivalent to the *amount* of potash which passes from the sulphuric to the nitric acid. If we then determine the quantities of lime and potash respectively which unite with a given weight of sulphuric acid we know that some other definite weight of nitric (or any other) acid will neutralize just these same quantities of both bases, so that when this quantity of acid has been determined experimentally for one base, it will hold good for the other also. Richter then prepared a large number of tables showing the quantities of different bases which would neutralize 1000 parts of a given acid, and parts of different acids which would neutralize 1000 of a given base, and was able to show that in the tables for different acids the bases all came in the same order and that the numbers involved were proportional. The same held true of the acids in the tables for bases.

Fischer.—In his books Richter confused these valuable ideas with fanciful speculations of no importance; his language was archaic, and his style such as might have been expected from a man who would invent the word *stoichiometry.* For all these reasons his work made little impression at the time, and might have been forgotten altogether, had not G. E. Fischer, who translated into German the works of Berthollet, introduced into the *Statique Chimique* a note summing up Richter's conclusions. Fischer was not only fortunate in his mode of expression, but he much improved the usefulness of Richter's tables by condensing them all into one. If to the table containing the quantities of bases which neutralize 1000 parts of sulphuric acid we add the quantities of acids having the same neutralizing power as 1000 parts of sulphuric acid, then the resulting table is a complete list of the weights of all acids and bases which are chemically equivalent. Fischer's table comprised twenty-one acids and bases. Some figures selected from it appear below, Fischer's original figures being compared with what would now be considered the correct ones. In calculating the latter the reader will bear in mind that the bases are computed as oxides and the acids as anhydrides, according to the theory then current which regarded salts as addition products.

Bases		True value	Acids		True value
Alumina	525	425	Sulphuric	1,000	1,000
Magnesia	615	503	Carbonic	577	550
Lime	793	700	Oxalic	755	900
Potash	1,605	1,180	Phosphoric	979	888
Soda	859	775	Nitric	1,405	1,350

Equivalency.—Neither Richter nor Fischer used the word, but we realize clearly that, so far as it goes, this is a true *table of chemical equivalents* or proportions by weight in which substances combine. It was the first of its kind. We can see readily enough that the idea might have been carried much farther without introducing any new principle. Richter recognized clearly that when a metal precipitates another the quantities involved are chemically equivalent, so that through a single quantitative relationship between one of the acids and one of the metals all of the latter might have been brought into the table. Indeed it could have been enlarged in this way until it included every element. We should then have a complete table of combining weights in which the numbers involved, would it is true, bear no superficial resemblance to our modern atomic weights; nevertheless, multiplication by a single factor would transform this table into one in which every number stood in a simple relation to those weights. Such a table would suffice for every stoichiometrical calculation.

Most of us have arrived at our conception of combining weights through the atomic theory and the gas-laws. It is therefore interesting to see that these relationships not only can be, but, as a matter of historical fact, actually were derived from a study of neutralization phenomena, without any use of the atomic conception and without any hypothesis concerning the ultimate constitution of matter.

Literature

BERTHOLLET'S *Essai de Statique Chimique* was published in 1803. The author's earlier paper delivered at Cairo is to be found in OSTWALD'S *Klassiker* No. 74.

A comprehensive treatment of the development of the science from the time of Lavoisier down to the date of its publication is to be found in KOPP'S *Die Entwickelung der Chemie in der Neueren Zeit*, Munich, 1873.

An excellent presentation of the same period is given in LADENBURG'S *Vorträge über die Entwickelungsgeschichte der Chemie*. An English translation of one of the earlier editions by DOBBIN has been published by the Alembic Club.

Claude Louis Berthollet
1748–1822

John Dalton
1766–1844

(Facing page 66)

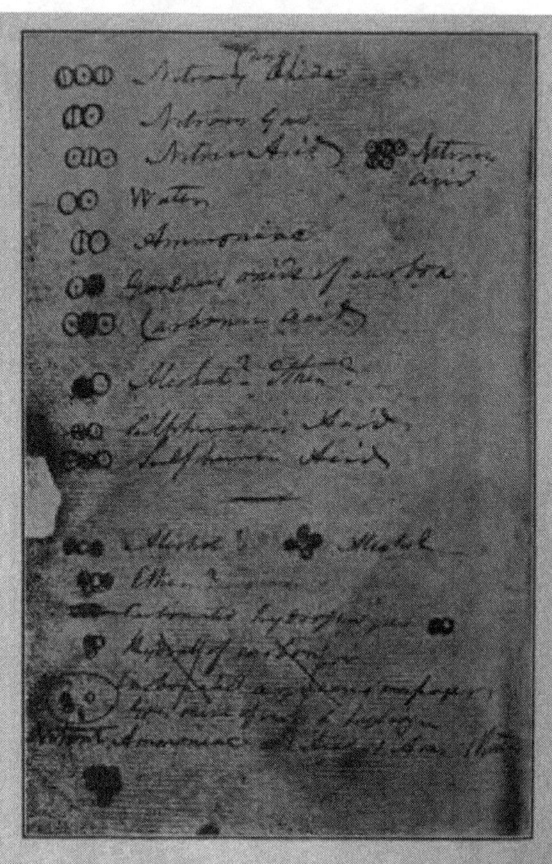

A PAGE FROM DALTON'S NOTEBOOK

Reproduced from Roscoe and Harden's "A New View of Dalton's Atomic Theory" by the kind permission of MacMillan and Co.

CHAPTER VIII

DALTON AND THE ATOMIC THEORY

he Atomic Conception.—The tendency to refer all physical nomena to the mechanical motions of corpuscles, atoms or mate particles is a very old one which helps us to visualize cesses whose details it is difficult to keep before the mind nout it. We have seen how this human tendency can be ed back at least as far as Democritus, and doubtless he was the first. All the physical sciences have used the conception ly at one time or another, and it took very definite form in mind of Newton who expressed himself on the subject as ows:

It seems probable to me that God in the beginning formed matter olid, massy, hard, impenetrable, movable particles, of such sizes figures, and with such other properties, and in such proportion, as it conduced to the end for which He formed them; and that these nitive particles, being solids, are incomparably harder than any ous bodies compounded of them; even so very hard as never to wear oreak in pieces, no ordinary power being able to divide what God nself made one in the first creation. While the particles continue re they may compose bodies of one and the same nature and texture ll ages; but should they wear away, or break in pieces, the nature hings depending on them would be changed."

The above is an admirable restatement of the ideas of Democ-us as applied to physics, and in a vague philosophical way to mistry, but no one before 1800 had realized that these same as might be so developed that they could give quantitative ount of the composition of all substances, and of their reactions n each other. This is what chemists understand by the mic Theory, and its discoverer was John Dalton.

Dalton.—Dalton once had occasion to write a brief account his own life and as it is short and characteristic it may be en entire:

"The writer of this was born at the Village of Eaglefield about 2 miles west of Cockermouth, Cumberland. Attended the village school there & in the neighborhood till 11 years of age, at which period he had gone through a course of Mensuration, Surveying, Navigation, &c., began about 12 to teach the Village School & continued 2 years afterwards; was occasionally employed in husbandry for a year or more; removed to Kendal at 15 years of age as assistant in a boarding School, remained in that capacity for 3 or 4 years, then undertook the same School as a principal & continued it for 8 years, & while at Kendal employed his leisure in studying Latin, Greek, French & the Mathematics with Natural Philosophy, removed thence to Manchester in 1793, as Tutor in Mathematics & Natural Philosophy in the New College, was 6 years in that Engagement, & afterwards was employed as private & sometimes public Instructor in various branches of Mathematics, Natural Philosophy & Chemistry chiefly in Manchester, but occasionally by invitation in other places, namely London, Edinburgh, Glasgow, Brimingham & Leeds."

There need only be added that Dalton was born about September 6, 1766, and that he died in Manchester, July 27, 1844. He was the son of a poor weaver, and as the little autobiography shows, he was from his earliest youth thrown upon his own resources. Yet with endless perseverance he always devoted every free hour to intellectual pursuits. He was a reserved silent man, frugal of words as of money (though kindly and generous in the essentials), and so regular in his habits that his neighbors could set their clocks by his movements—a man distrustful of the results of others, who had to work everything out in his own original way. He was by no means gifted as an experimenter, and financial reasons long made it necessary for him to work with the crudest apparatus, often of his own construction; yet all these handicaps could not prevent him from discovering several of the laws which rest at the very foundation of the science.

Color Blindness.—It is interesting to see that he accomplished all this by taking up the work which lay nearest, beginning with studies of his own color-blindness and the weather. Dalton was almost the first to make a systematic study of color-blindness, and though his investigation yielded no important results it showed all his characteristic diligence. Dalton's own case must have been an extreme one, for in describing one of his lectures at the Royal Institution he wrote to a friend: "In lecturing on optics I got six ribands—blue, pink, lilac, and red, green and brown—which matched very well, and told the curious audience so."

Studies in Meteorology.—No subject seems to have interested Dalton so much as meteorology. He faithfully kept a daily record of the weather and allied phenomena from 1787 till the very day before his death, and it contains no less than two hundred thousand separate observations. The brief vacations which he allowed himself were spent mostly in the Lake District where it was his favorite occupation to climb mountains with such instruments as he could carry, in order to compare atmospheric conditions at different altitudes. All this led naturally to an interest in gases and a study of their properties in the laboratory. In the first year of the nineteenth century we find him reading a paper before the Manchester Literary and Philosophical Society in which he announces the discovery of some of the most important laws concerning gases. Among these were the law that gases expand equally for a given rise of temperature, that the vapor pressures of liquids are the same at equal intervals of temperature above and below their boiling points, that at constant volume each gas in a mixture exerts the same pressure as if the other gases were absent, and consequently, that the solubilities of mixed gases are proportional to their partial pressures. Dalton had also observed that the composition of the atmosphere was independent of the altitude, and he had shown that this was not due simply to mechanical agitation, by experiments in which heavy gases diffused upward into lighter ones, while the latter diffused downward, even through very narrow tubes.

The Atomic Theory.—The atomic theory grew more naturally out of studies of this kind than had been realized until comparatively recently. It was not published by Dalton till his *New System of Chemical Philosophy* appeared in 1808, but he communicated the fundamental ideas involved informally some years earlier, the first occasion being a table of atomic weights which is appended practically without comment to a paper delivered in 1803 on the subject of the solubility of mixed gases. The late Sir Henry Roscoe did a valuable service by unearthing from the archives of the Manchester Literary and Philosophical

Society, Dalton's laboratory notebooks and other manuscripts which now make it clear just how the conception grew up in his mind. One of the most important of these documents is the syllabus of a lecture delivered in 1810 which reads in part as follows:

"Having been long accustomed to make meteorological observations, and to speculate upon the nature and constitution of the atmosphere, it often struck me with wonder how a *compound* atmosphere, or a mixture of two or more elastic fluids, should constitute apparently a homogeneous mass, or one in all mechanical relations agreeing with a simple atmosphere.

"Newton had demonstrated clearly in the 23d Prop. of Book 11 of the *Principia* that an elastic fluid is constituted of small particles or atoms of matter which repel each other by a force increasing in proportion as their distance diminishes. But modern discoveries having ascertained that the atmosphere contains three or more elastic fluids of different specific gravities, it did not appear to me how this proposition of Newton's would apply to a case of which he, of course, could have no idea. The same difficulty occurred to Dr. Priestley, who discovered this compound nature of the atmosphere. He could not conceive why the oxygen gas, being specifically heaviest, should not form a distinct *stratum* of air at the bottom of the atmosphere, and the azotic gas[1] one at the top of the atmosphere. Some chemists upon the Continent—I believe the French—found a solution of the difficulty (as they apprehended). It was *chemical affinity*. One species of gas was held in solution by the other; and this compound in its turn dissolved water—hence *evaporation*, rain, etc. This opinion of air dissolving water had long been the prevailing one, and naturally paved the way for the reception of that which followed—of one kind of air dissolving another. It was objected that there was no decisive *marks* of chemical union when one kind of air was mixed with another. The answer was, that the affinity was of a very slight kind, not of that energetic cast that is observable in most other cases. I may add, by-the-bye, that this is now, or has been till lately, I believe, the prevailing doctrine in most of the chemical schools in Europe. In order to reconcile—or, rather, adapt—this chemical theory of the atmosphere to the Newtonian doctrine of repulsive atoms or particles, I set to work to combine my atoms upon paper. I took an atom of water, another of oxygen, and another of azote, brought them together and threw around them an

[1] Nitrogen.

atmosphere of heat as per diagram[1] I repeated the operation, but soon found that the watery particles were exhausted (for they make but a small part of the atmosphere). I next combined my atoms of oxygen and azote one to one; but I found in time my oxygen failed. I then threw all the remaining particles of azote into the mixture, and began to consider how the general equilibrium was to be obtained. My triple compound of *water*, *oxygen*, and *azote* were wonderfully inclined, by their superior gravity, to descend and take the lowest place. The double compounds of *oxygen* and *azote* affected to take a middle station; and the azote was inclined to swim at the top. I remedied this defect by lengthening the wings of my heavy particles—that is, by throwing more heat around them, by means of which I could make them float in any part of the vessel. But this change, unfortunately made the whole mixture of the same specific gravity as azotic gas. This circumstance would not for a moment be tolerated. In short, I was obliged to abandon the hypothesis of the chemical constitution of the atmosphere altogether as irreconcilable to the phenomena. There was but one alternative left—namely, to surround every individual particle of *water*, of *oxygen*, and of *azote* with heat, and to make them respectively centres of repulsion, the same in a *mixed* state as in a *simple* state. This hypothesis was equally pressed with difficulties, for still my oxygen would take the lowest place, my azote the next, and my steam would swim upon the top. In 1801 I hit upon an hypothesis which completely obviated these difficulties. According to this, we were to suppose that atoms of one kind did *not* repel the atoms of another kind, but only those of their own kind. This hypothesis most effectually provided for the diffusion of any one gas through another, whatever might be their specific gravities, and perfectly reconciled any mixture of gases to the Newtonian theorem. Every atom of both or all the gases in the mixture was the centre of repulsion to the proximate particles of its own kind, disregarding those of the other kind. All the gases united their efforts in counteracting the pressure of the atmosphere, or any other pressure that might be exposed to them.

"This hypothesis, however beautiful might be its application, had some improbable features. We were to suppose as many distinct *kinds* of repulsive powers as of gases; and, moreover, to suppose that *heat* was not the repulsive power in any one case—positions certainly not very probable. Besides, I found from a train of expts. which have been published in the 'Manchester Memoirs' that the diffusion of gases through each other was a *slow* process, and appeared to be a work of considerable effort.

[1] The diagrams referred to in this quotation are no longer accessible.

"Upon considering this subject, it occurred to me that I had never contemplated the effect of *difference of size* in the particles of elastic fluids. By *size* I mean the hard particle at the centre and the atmosphere of heat taken together. If, for instance, there be not exactly the same *number* of atoms of oxygen in a given volume of air as of azote in the same volume, then the *sizes* of the particles of oxygen must be different from those of the azote. And if the *sizes* be different, then—on the supposition that the repulsive power is heat—no equilibrium can be established by particles of unequal sizes pressing against each other. (*See* diagram.)

"This idea occurred to me in 1805.[1] I soon found that the *sizes* of the particles of elastic fluids *must* be different. For a measure of azotic gas and one of oxygen if chemically united, would make nearly *two* measures of nitrous gas, and these *two* could not have *more* atoms of nitrous gas than the *one* measure had of azotic or oxygen. (*See* diagram.) Hence the suggestion that all gases of different kinds have a difference in the *size* of their atoms; and thus we arrive at the reason for that diffusion of every gas through every other gas, without calling in any other repulsive force than the well-known one of heat. This, then, is the present view which I have of the constitution of a mixture of elastic fluids."

* * * * * * * * * * * *

"The different *sizes* of the particles of elastic fluids under like circumstances of temperature and pressure being once established, it became an object to determine the relative *sizes* and *weights*, together with the relative *number*, of atoms in a given volume. This led the way to the combinations of gases, and to the number of atoms entering into such combinations the particulars of which will be detailed more at large in the sequel. Other bodies besides elastic fluids—namely, liquids and solids—were subject to investigation, in consequence of their combining with elastic fluids. Thus a train of investigation was laid for determining the *number* and *weight* of all chemical elementary principles which enter into any sort of combination one with another."

Dalton's Reasoning.—Nothing could show more clearly than the above, the mechanical trend of Dalton's mind and the pictorial methods by which he reached his conclusions. It is not chemical analysis as with Richter which leads him through the law of multiple proportions to the conception of atoms. On

[1] Roscoe points out that Dalton must have mistaken the date here, for he had given a table of atomic weights in 1803.

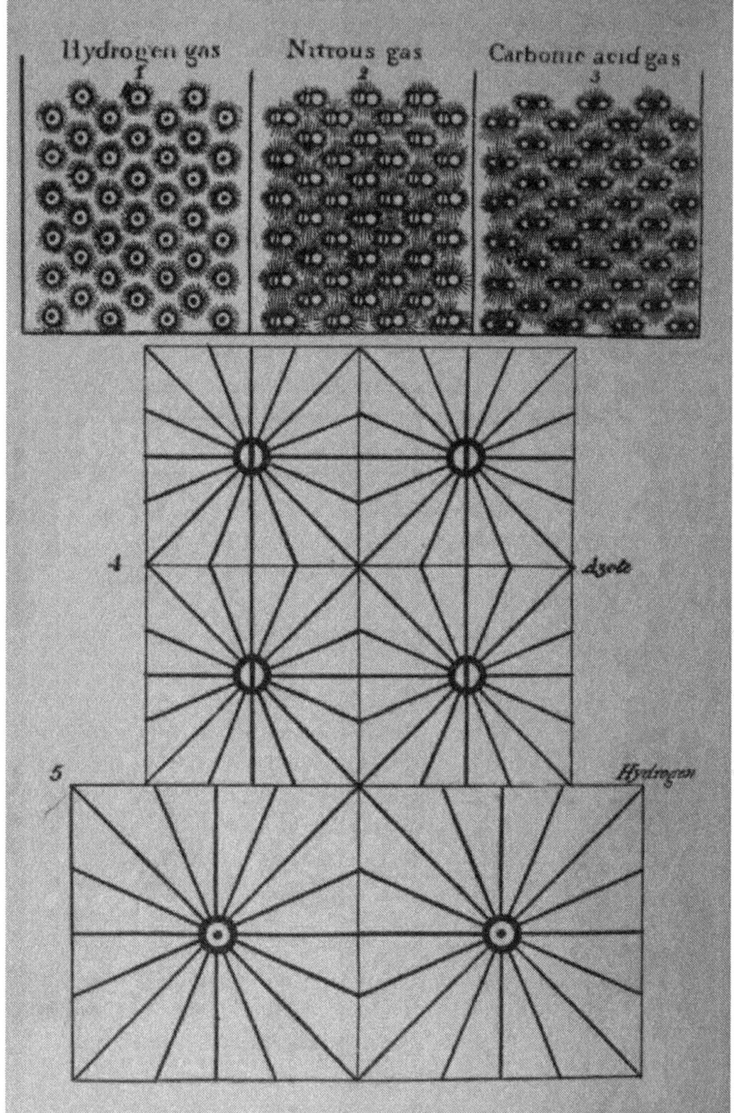

Some of Dalton's Pictures of Atoms

(Note the *different volumes* occupied by atoms of hydrogen and nitrogen.)
Reproduced from Roscoe and Harden's "A New View of Dalton's Atomic Theory" by the kind permission of MacMillan and Co.

the contrary the atomic hypothesis is always a part of the working machinery of his mind. He solves his problems by putting the individual particles down on paper and patching them together. The diagrams in his *New System* bring this out most strongly. Dalton depicts his hard particle in the center of a square and fills in the intervening space with rays emanating from the atom. These constitute his "atmosphere of heat," and the whole looks much like a spider. Finally he ranges his atoms of one gas over those of another, and seeing that squares of equal size would fit over each other perfectly, each ray meeting a corresponding ray from another atom, he takes this as conclusive evidence that gases whose atoms were of the same size could not diffuse through each other. If however they were of different sizes, then the rays would meet unevenly, motion would be set up, and the gases would interpenetrate. Knowing the density of the different gases Dalton now sees that he could calculate the relative diameters of his atoms if he knew their relative weights, so he proceeds to study the best available analyses and to make others of his own in order to determine the proportion by weight in which the elements combine.

The Law of Multiple Proportions.—As he anticipates, this reveals the law of multiple proportions, namely, that the weights of an element which unites in more than one proportion with another element stand to each other in a simple ratio. This he takes as a confirmation of his original hypothesis and, adopting hydrogen as a standard of reference, he proceeds to calculate his atomic weights. Here, however, he encounters difficulties, the seriousness of which he seems at first to have hardly realized.

Dalton's Atomic Weights.—In Dalton's time water was the only known compound of oxygen and hydrogen. The analysis tells us that in this substance the weight of oxygen is eight times that of hydrogen. The atomic weight which we select for oxygen, however, will vary according to the number of atoms which we accept as entering into combination in this substance. If the formula is HO_8 then the atoms of oxygen have the same weight as those of hydrogen; if it is HO_{16} then they are only half as heavy and so on. Dalton had no data by which he could decide such a question, yet realizing the immense practical utility of his discovery he felt that it justified some assumption and he selected the simplest. Where the elements united in but one proportion he assumed that but one atom of each was concerned. When they united in more than one proportion the compound which was best known received the simplest formula, this of course being modified by such considerations as made the atomic weight consistent when derived from two or more compounds of the same element. Dalton accordingly adopted the formula HO for water, which gave oxygen an atomic weight of 8. Similarly, since ammonia was then the only known compound of hydrogen and nitrogen, he gave it the formula NH, from which we should now calculate the atomic weight of nitrogen as 4.5. Similarly carbon would be 6 if ethylene were CH and marsh-gas CH_2. Dalton's own figures differ widely from these but the discrepancy is due to the imperfections of quantitative analysis at that time. His symbols were original with him, and he never gave them up, indeed he characterized our modern alphabetical ones as "unscientific!"

Objections to the Theory.—Richter's work had not been widely read, and the ideas of Dalton so simplified chemical thinking and calculation that they were hailed with great enthusiasm. The more clear-sighted, however, soon realized that his assumption which assigned the simplest formula to the best-known compound had no rational foundation. They pointed out that there was no known criterion by which the number of atoms taking part in a given combination could be determined, and while they acknowledged the value of the law of multiple proportions they felt that, stripped of its speculative superstructure, Dalton's discovery amounted to simply this: that the elements always combine in proportions by weight which are multiples of a certain unit. These men recognized, of course, that this law could not be visualized except on the assumption that combination took place by atoms, and that the relative weights of such atoms must be simple multipla or submultipla of the combining weights obtained by Dalton or Richter. What they denied was that there was any way of determining *which* multipla ought to be selected. Others had more faith. They saw that the atomic hypothesis, if true, gave a deeper insight into the constitution of matter than could be obtained without it, and while they realized that many of Dalton's numerical values must be wrong, they trusted to the future for the dicovery of data which should justify his fundamental idea.

Gay-Lussac.—In 1808, the same year in which Dalton published the first portion of his *New System*, important data bearing directly upon the subject of the atomic weights were furnished by Joseph Louis Gay-Lussac. This brilliant scientist was born at St. Leonard in 1778. He had received an excellent education in Paris, and had become an assistant of Berthollet who admitted him to his celebrated *Société d'Arceuil.* He was destined to high distinction; he became professor at the *Sorbonne* and later at the *Jardin des Plantes* and in 1850, the year of his death, he was made a peer. At the time we are discussing he was at the outset of his scientific career, which had opened in 1802 with a paper on the effect of temperature upon the volumes of gases. In this he enunciated the famous law now generally known by his name, to the effect that for every degree of temperature all gases expand $1/273$ of their volume at 0°. He ascribed the priority, however, to Charles.

Chemical Combination by Volume.—In 1805 Gay-Lussac and Alexander von Humbolt had collaborated in an investigation intended to determine the exact proportions by volume in which oxygen and hydrogen combine under the influence of the electric spark. They found that no matter which gas was in excess 100 measures of oxygen always reacted with just 200 measures of hydrogen within the limits of experimental error. Struck with the simplicity of this relation Gay-Lussac continued the investigation to see whether similar relations obtained in the case of other gases. He found this to be the case, and in 1808 he was able to show, among other things, that 100 measures of ammonia gas unite with just 100 of hydrochloric acid; that 100 measures of nitrogen unite with 50 of oxygen to form nitrous oxide, with 100 to form nitric oxide, and with 200 to form nitrogen peroxide; that ammonia is formed by the union of one volume of nitrogen with three of hydrogen, and that the volumes of oxygen in the two oxides of sulphur stand in the ratio of two to three. Gay-Lussac expressed his conclusions in the modest form that chemical reactions between gases take place in simple volume ratios, and that when contraction occurs in such reactions the diminution in volume stands in a simple ratio to the volumes of the original gases.

Bearing upon the Atomic Weights.—Since chemical reactions take place between simple multiples of the combining weights, and since the volumes of combining gases also stand in a simple ratio it follows that the weights of equal volumes must stand in a simple ratio to the combining weights. In gases the space between the atoms must be large in proportion to that occupied by the atoms themselves, so it did not seem unreasonable to suppose that the volume (sphere of influence) occupied by an atom might be the same for all gases, and this would lead to the conclusion that the weights of equal volumes are strictly proportional to the atomic weights. Gay-Lussac makes no attempt to revise Dalton's atomic weights on this basis, but he does say that he considers his own work a valuable confirmation of Dalton's fundamental idea, and that it offers a basis for the selection of multipla less arbitrary than Dalton's original assumptions. This view of the case appealed to many others and was urged upon Dalton by his friend Professor Thomson of Glasgow and by Berzelius.

Dalton's Attitude.—Dalton, however, declined to make any use of the idea, partly, no doubt, because, as we have seen, the atomic theory had originally been suggested to his mind by the belief that the particles of gases are *not* of the same size. In 1812 he writes to Berzelius:

"The French doctrine of *equal measures* of gases combining, etc., is what I do not admit, understanding it in the mathematical sense. At the same time I acknowledge there is something wonderful in the frequency of the approximation."

And in his manuscript notes for a lecture delivered in 1807 appears the following:

"*Query*, are there the same *number* of particles of any elastic fluid in a given volume and under a given pressure? No; azotic and oxygen gases mixed equal measures give half the number of particles of nitrous gas, nearly in the same volume."

Dalton's "mathematical" objection arose from the fact that he quite erroneously supposed Gay-Lussac's experimental results to be less trustworthy than some of his own and of his

Joseph Louis Gay-Lussac
1778–1850

(Facing page 78)

Amadeo Avogadro
1776-1856

riend Henry which gave less simple ratios. The second argument discloses a real difficulty which remained insuperable to many minds at that time. Nitric oxide *does* occupy the same volume as the constituent quantities of nitrogen and oxygen which compose it, and yet it is clear that the number of "atoms" of nitric oxide must be one-half that of the oxygen and nitrogen combined; or, as we shall say today, there are but half as many *molecules* of nitric oxide as there are *atoms* of nitrogen and oxygen which combine to form it. As we know, many other gas reactions present the same difficulty.

Avogadro's Hypothesis.—A solution was found by Amadeo Avogadro (1776–1856), long professor of physics at Turin. In 1811 he published in the *Journal de Physique* an article setting forth the hypothesis which we now connect so prominently with his name. In it he shows that such discrepancies as that just mentioned can be harmonized if we assume that the smallest particles of the elementary gases thought of by Dalton are themselves compound, just as in the so-called compound gases. He draws a distinction, therefore, between *molecules intégrantes* (the *physical* units of gases which determine their volume, etc. and for which we still retain the name molecule), and *molecules élémentaires* (the *chemical* units for which we now use the word atom). Of only the former is it true that equal volumes contain an equal number. He points out that any convenient number of atoms may be assumed in the molecule of an elementary gas, though for the common ones two is the maximum hitherto found necessary, and he shows that reactions formerly interpreted as cases of simple addition must now be considered as involving decomposition followed by addition or metathesis; so that in the special case just mentioned we can write as we do today:

$$N_2 + O_2 = 2NO$$

In 1814, the eminent physicist Ampère advanced similar ideas, though he did not put the case quite as clearly as Avogadro had done, for he assumed four instead of two atoms in the molecules of the elementary gases, and he confused the issue by a fruitless attempt to connect the atomic constitution of solids with their crystalline form.

Reception of the Theory.—These ideas of Avogad Ampère play so important a part in modern chemical that it is difficult to realize that, when first published, the no impression whatever.

In the sequel we shall often feel that much unfortuna fusion might have been avoided if Dalton and Gay-Lussa have adopted the suggestion of Avogadro and made c cause to present these views to the world. There is a that, in such a case, the atomic and molecular hypothese have taken, far earlier than they did, the place in cl thinking which they hold today. What might have b however, as fruitless a topic in history as in the more f relations of common life, and we can only ask why the i Avogadro received so little attention from his contemp

It has been suggested that this was because Avogad visionary and accompanied his theoretical speculations experimental work. We also know that some chem eminence impatiently denounced his views on the groun they involved fractions of atoms, whereas the word atom something which cannot be divided (*Greek* α-primitive an to cut). We cannot seriously believe, however, that e the earlier days of the nineteenth century, scientific men reject a really useful hypothesis purely on grounds of etyn The real reason lies deeper. The scientific world judg potheses by certain hard standards which are on the whc and sound. These demand that a hypothesis shall no explain the given series of facts for which it was ori designed, but that it shall also give account of others remote, and shall lead to the discovery of new facts a lationships, in short that it shall enable us to make predi At this time Avogadro's hypothesis was not in a position these things and therefore it was justly (even if unfortu disregarded. As a result, the atomic theory failed to com to its own for fifty years.

Wollaston's Equivalents.—In England especially, a po sceptical school soon sprang up, of which the most imp representative was William Hyde Wollaston (1766–1828). aston made some analyses of neutral and acid salts brilliantly confirmed the law of multiple proportions,

vertheless refused to accept the atomic theory, essentially for
; reasons already stated on page 76. He fully recognized,
wever, the practical value of the combining weights and in
14 we find him busy determining what he calls *equivalents.*
ollaston was the first to use this word in a chemical sense, and
is deserves more than passing notice because its meaning has
dergone important changes in the course of the many contro-
rsies in which it has figured since his day. Now, of course,
has a rather restricted meaning, and refers to such weights of
gents as balance or neutralize each other, as in the case of acids
d bases, or of metals which precipitate each other. The
ative weights determined by Richter were, therefore, true
iivalents in the modern sense. Wollaston's equivalents, how-
er, were only multipla or submultipla of the combining weights,
ected solely on the basis of maximum simplicity and con-
tency. They were therefore, in principle, nothing else than
lton's atomic weights without their speculative significance.

Prout's Hypothesis.—If the atomic theory had opponents it
o found some friends whose enthusiasm was not always tem-
red by discretion. Among these was William Prout (1785–
50), an English physician much interested in physiological
emistry who, in 1815, published an anonymous paper in which
called attention to the closeness with which the atomic weights
the elements so far as then determined approximated whole
mbers. He did not stop there but expressed the opinion that
drogen was therefore the universal substance, and that the
oms of other elements were really aggregates of hydrogen atoms.
out soon made eminent converts, including Dalton's friend,
omson, and this doubtless because he gave expression to the
iversal repugnance which we all feel toward believing in
venty or more different kinds of matter. In its original form
e hypothesis was easy to disprove as soon as the atomic weight
chlorine, for example, was fixed at approximately 35.5, but
more ardent supporters of course immediately proceeded to
luce their hypothetical unit first to the half, and then to the
th of an atom of hydrogen; always keeping a little within the
creasing experimental error. We are, of course, not concerned
th discussions of this kind. What interests us historically
the fact that the fundamental idea underlying Prout's hypothe-

sis, from the *materia prima* of the alchemists to the electrons of the present day, is something which will not down, and which has had much influence upon the work and thinking of many master-minds.

Literature

DALTON'S *New System of Chemical Philosophy* was published in 1808. There is an excellent life of Dalton by his friend HENRY, published by the Cavendish Society in 1854. There is a more recent one by SIR HENRY ROSCOE, entitled *John Dalton and the Rise of Modern Chemistry*, 1895. A book which deserves special attention is ROSCOE and HARDEN'S *A New View of the Origin of Dalton's Atomic Theory*, 1896.

Contributions of Dalton, Gay-Lussac, Avogadro, and Wollaston to the atomic and molecular theories are to be found in the *Alembic Club Reprints* Nos. 2 and 4. See also OSTWALD'S *Klassiker* No. 3 for the work of Dalton and Wollaston, and No. 8 for that of Avogadro and Ampère.

Luigi Galvani
1737-1798

ome of Galvani's Experiments with Frogs' Legs as Depicted in his Original Communication

(Facing page 82

ALESSANDRO VOLTA
1745-1827

CHAPTER IX

HE EARLY HISTORY OF GALVANIC ELECTRICITY

/hile men like Ritcher and Dalton were making their great tributions to pure chemistry, Italian scientists were studying manifestations of a new form of energy. This was galvanic tricity—something hitherto entirely overlooked—but destined ceforward to exert the greatest influence not only upon chemy but in almost every department of life.

alvani's Discovery.—A paper published in 1791 by Luigi vani, a distinguished physician of Bologna, describes the accit which first drew his attention to the subject. He relates he was one day in his laboratory where some partially dised frogs were lying on a table near a static electrical machine. hen so happened that one of his assistants touched the bare ral nerve of one of the frogs with a scalpel just at the moment n a spark was drawn from the machine, and was surprised notice a sharp twitching of the frog's leg. Galvani's atten- was called to the occurrence and a series of systematic eximents was immediately begun. These showed that the effect ld be produced at will so long as the nerve was touched with metal, and that the twitching was more violent if at the e time the frog's leg was also connected with the ground. e the twitching was induced by the electric spark and did take place when the nerve was touched with a non-contor, Galvani concluded that the phenomenon was electrical tried numerous experiments to see if it could be induced atmospheric electricity in thunder storms. In the course these experiments, another accident due to arranging the s upon hooks showed that neither the spark of the electric hine nor lightning in the vicinity were necessary to proe the effect. It was sufficient to put the crural nerve and extremity of the leg in metallic contact, and the action much stronger if the circuit connecting them consisted of two different metals. Furthermore, the nature of the two metals made a perceptible difference in the violence of the muscular contraction. In his interpretation Galvani does not seem to have laid as much weight on this fact as he should have done. The relationship between the nerve and muscle suggested to him an analogy with the Leyden jar, and he confirmed this to his satisfaction by coating both nerve and muscle with tin foil and thereby accentuating the effect. This led Galvani to name the new force animal electricity and to believe that it had its source in the organs of animals, where it might play an important rôle in physiological processes.

Volta's Explanation.—Galvani's paper produced a great impression and numerous scientists repeated the experiments. Among these was Alessandro Volta (1745–1827), professor of physics in Pavia, whose conclusions, however, differed radically from those of Galvani. In the first place, he laid weight upon the essential condition that the ends of the metallic circuit must be different in order to produce the twitching. They may indeed be of the same metal, but in this case there must be at least some difference in the surfaces, or the effect will not be shown. Further, he produced the same effect by connecting two points on the nerve by a metallic circuit in which the muscle was not included, and in this way showed that the muscle was not essential to the phenomenon; it therefore formed no part of a Leyden jar, and its contraction was simply a secondary effect due to the irritation of the nerve. Considering then the nerve and the two metals, it seemed reasonable to suppose that the former might be replaced by any moist conductor, and it was soon found that this was the case. As early as 1760 J. G. Sulzer had called attention to the fact that when the moist tongue is thrust between two plates of different metals which remain in contact at their edges a peculiar taste is observed, and it is clear that this sensation, like the twitching of the frog's leg, is merely an extremely sensitive electroscope which permits the detection of currents far weaker than any which had hitherto been studied by such apparatus as was then in use—all of which had been designed for dealing with the high potentials of static electricity. Volta conducted many series of experiments in which the metals and moist conductors were all varied and made every effort in his power to characterize the new phenomena as truly electrical. At last, by means of a condenser of his own design he succeeded in multiplying the charge until it would show the familiar effect upon an electroscope, and rightly concluded that the essential difference from static electricity was greater quantity at a lower potential. As a result of these studies Volta utterly rejected the animal origin which Galvani had ascribed to the phenomena he had observed and sought the cause rather in the mere contact of the two metals. Volta next tried to determine what electrical relation dry metals had to each other. He brought them in contact, then separated them, and by tests with an electrometer satisfied himself that whenever two metals come into proximity one exhibits a positive and the other a negative charge. Even as early as 1792 Volta published a list of substances in such an order that each is positive toward all which follow it and negative to all which precede. This was the first *potential series*. By its aid one should always be able to predict the direction of the current when any two metals in contact are also separated by a moist conductor.

The Contact Theory.—Volta's theoretical conception of the matter was extremely simple. Every solid contains the "electric fluid" under a state of tension characteristic for that solid. If now two solids are brought into contact the fluid passes from the region of the higher to that of the lower tension. If in addition there is a moist conductor in the circuit this acts—'in a manner not yet thoroughly understood'—as a kind of 'semi-permeable wall' through which the fluid may pass back to its original source. This was Volta's celebrated contact theory of electricity. Its author fully realized that it involved perpetual motion, but our modern ideas concerning the conservation of energy had little place in the scientific thinking of the day, and Volta was rather proud than otherwise of this aspect of the case. He had some right to be proud of his theory, for in spite of this fundamental fault, it was so simple and self-consistent that it held the field as the most practicable working hypothesis for several decades.

The Chemical Theory.—This contact theory of Volta had hardly been formally stated before another theory essentially chemical was set up in opposition by J. W. Ritter (1776–1810).

Ritter called attention to the fact that when two metals in moist condition are left in contact, corrosion of one of them proceeds far more rapidly than when they are isolated. He rightly interpreted this as an electrical phenomenon and ascribed the current to the oxidation of one of the metals. He contended that without the oxidation there is no current, and found a convincing argument in the fact that Volta's potential series might

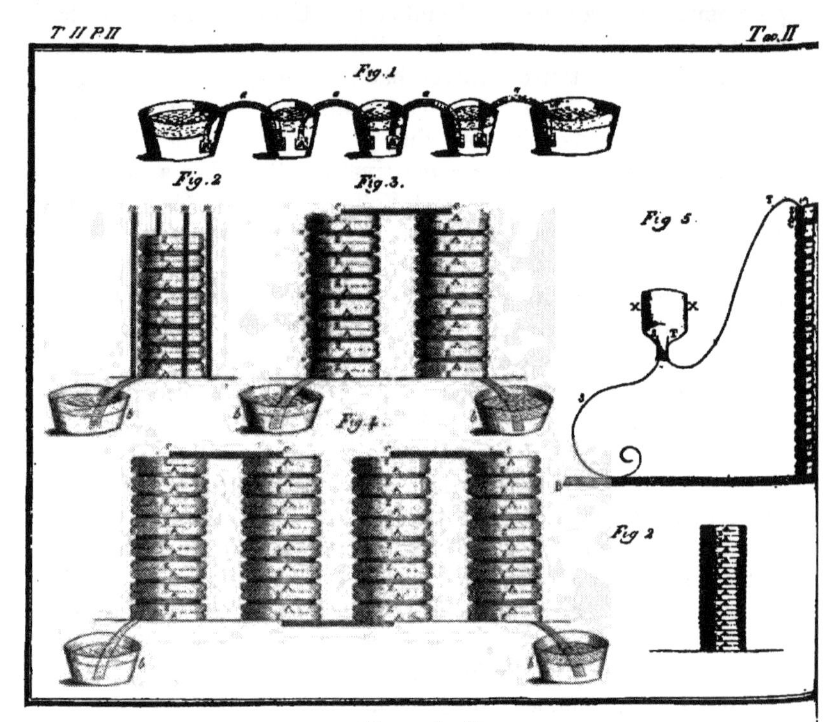

VOLTA'S PILE

serve equally well as a list representing the order of the relative affinities of the substances concerned for oxygen.

Volta's Pile.—Volta meanwhile was busy with attempts to obtain stronger currents[1] and attained this object by a logical

[1] In 1792 Valli ingeniously attempted to accomplish a similar result by connecting the nerves and legs of fourteen frogs in series. The results were ambiguous.

application of his contact theory. He reasoned essentially as follows: The electric fluid passes from silver to zinc and if a moist conductor is present it can pass through this to silver again. Hence if we add a second pair of plates in the same order the second silver passes the fluid on to the second zinc with its own intensity augmented by that of the current which it has received from the moist conductor. Volta accordingly laid upon a plate of silver one of zinc, and upon this a layer of cloth or pasteboard dipped in water or a salt solution, and then repeated the series indefinitely, getting stronger and stronger currents as the number of plates increased. Finally, he obtained not only direct effects upon the electroscope, but also the familiar static phenomena of shock and spark. This was the famous *pile* of Volta. He described it in a letter addressed to Sir Joseph Banks dated March 20, 1800.

The First Electrolysis.—The communication excited the greatest interest, and experiments with the new apparatus began in almost every laboratory. All were struck with the remarkable chemical effects produced. In May of the same year Nicholson and Carlisle described the decomposition of water by the current and solutions of numerous salts, acids and bases were soon after subjected to electrolysis. It is interesting that many phenomena which now seem to us a matter of course then caused something akin to amazement. That water should be decomposed by the current was by no means unexpected, since Van Troostwijk and Deimann had already achieved that result in 1789 with the aid of a powerful static machine. What caused surprise was the fact that the oxygen and hydrogen appeared separately. Where was the water decomposed? If at the positive pole, for example, how did the hydrogen get across to the negative pole unobserved? Similar difficulties arose in the case of salt solutions. If potassium sulphate be electrolyzed, acid appears at one pole and alkali at the other. Even if the solution is strongly alkaline the liquid in the vicinity of the positive pole soon becomes acid. Experiments like the following were tried: The negative pole of a battery was surrounded by a potassium sulphate solution and the positive by water. Between the two and in communication with both was a vessel containing strong alkali. Sulphuric acid soon appeared at the positive pole. How did it get through all the alkali without being neutralized on the way?

Grotthuss's[1] Theory of Electrolysis.—Such puzzles proved very troublesome until Ch. J. D. von Grotthuss in 1805 explained the phenomena by a successive decomposition and recombination among the molecules of the electrolyte. In the case of water, for example, the negative pole attracts an atom of hydrogen from an adjacent water molecule. This is evolved as gas; and the oxygen left over robs another molecule of its hydrogen to take the place of the first. This goes on all the way to the other electrode where the last oxygen is attracted to the positive pole and is evolved in its turn. This explanation proved adequate for a long time.

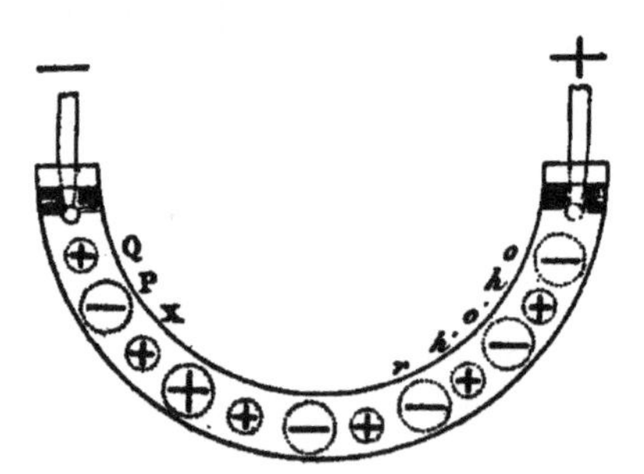

THE MECHANISM OF ELECTROLYSIS ACCORDING TO GROTTHUSS

The Contact and Chemical Theories in Opposition.—These discoveries strengthened the faith of those who believed in the chemical origin of the current, but Volta himself remained blind to this aspect of the question. In his first communication he had mentioned none of the chemical effects of his battery in spite of the frequent opportunities he must have had to observe them. He described, for example, putting both terminals into the same vessel of water but he said no word of gas evolution. Later, when such effects were called to his attention he expressed a good deal of surprise and interest, saying in substance that his battery was so wonderful a thing that remarkable effects both physical and chemical might well be expected of it, but he regarded

[1] The name appears in this form in the journals where most of his work is published. At least three other spellings are to be found in well-known modern books.

the latter as secondary and accidental. So the two schools stood opposed to each other as they were destined to do for years to come. The "chemists," as they came to be called, could always maintain that contact offers no explanation for the work done, that there never is a current without chemical action, and that when the chemical action ceases the current stops. The "physicists," on the other hand, found their chief argument in the potential existing between dry metals, but they could also hold up to their opponents that the strength of the current stands in no definite relation to the chemical reaction going on in the cell, that very vigorous chemical reactions often proceed without producing any electrical phenomena whatever, and finally that when such chemical action does take place, it is often demonstrable that it begins only when the circuit is closed and therefore cannot be the cause of the current. The details of this long controversy are extremely interesting but they belong to the special history of electrochemistry. Here we shall deal with that subject only where it exerts a marked influence upon the development of the science as a whole. The foregoing account of its beginnings, however, seems an appropriate and necessary introduction to any discussion of the work of Humphry Davy.

Literature

To the student of electrochemistry OSTWALD's *Elektrochemie, ihre Geschichte und Lehre* cannot be too highly recommended. With admirable clearness the author traces the whole development of the subject from the earliest observations of Galvani down to about 1895. The text consists largely of extensive quotations from the work of the original investigators unified and illuminated by clear and helpful running commentary. In spite of its formidable size the book is a model of what an intensive historical study should be.

GALVANI's original pamphlet is reprinted in OSTWALD's *Klassiker* No. 52, as are also the earlier papers by VOLTA, Nos. 114 and 118, and those by GROTTHUSS, No. 152.

CHAPTER X

HUMPHRY DAVY

Humphry Davy was the son of a wood-carver in Penzance, where he was born December 17, 1778. He was of a precocious turn but did badly at school on account of his extremely buoyant disposition and love of sport. Two of his early interests remained with him through life—his passion for fishing and for writing verse—indeed in later years some of his poetry earned a good deal of praise. On the death of his father Davy entered the office of a local physician whom he assisted in the preparation of remedies; but he soon developed such a taste for startling experiments and explosions that his employer saw him go without regret. One of his friends, however, who had heard of his interest in chemistry gave Davy an introduction to Dr. Thomas Beddoes which was destined to be a turning-point in his life. The discovery of new substances of striking properties always inspires physicians to try these upon their patients, and the recent investigations of Priestley and others had led Dr. Beddoes to found in Bristol what he called a "pneumatic institute." Here he intended to prepare the new gases and experiment upon their physiological action. Davy was put in charge of this laboratory in 1798, and immediately began preparing and inhaling gases. Some of these experiments nearly cost him his life, but his perseverance was soon rewarded by the discovery of the remarkable physiological action of nitrous oxide, now familiarly spoken of as laughing gas. This discovery made a popular appeal, and the inhalation of nitrous oxide became a fad, with the result that Davy soon acquired a popularity which doubtless won him the professorship at the Royal Institution in 1801.

The Royal Institution.—Not long before this time, Count Rumford, who was then very influential in the scientific circles of London, had persuaded some friends of kindred tastes to unite with him in establishing this institution. It exists for the

HUMPHRY DAVY
1778–1829

(Facing page 90)

A Séance at the Royal Institution in Davy's Time. (*From a Contemporary Cartoon.*)

Reproduced from Thorpe's "Humphry Davy—Poet and Philosopher" by the kind permission of Cassel and Co.

purpose of securing for those interested in such topics courses of lectures dealing with the latest discoveries in science and the arts. The professor in charge has a completely equipped laboratory at his disposal, primarily for the requirements of the lectures, but also for the prosecution of his own researches. The standard of appointments has always been kept high, so that from that day to this the professors have been scientists of great eminence, and the Institution has remained an important centre in English scientific life.

Davy was only twenty-two when he received the appointment but his lectures at once aroused enthusiasm on account of the brilliancy of his delivery and skill in experimentation. At the same time his personal magnetism made him immensely popular, so that fashionable society, which had smiled upon him in Bristol, fêted and lionized him in London. Davy was knighted in 1812 and made a baronet in 1818. The combined burdens of vigorous scientific and active social life, however, seriously impaired his health, and he broke down altogether in 1826. Repeated journeys to the Continent brought little relief, and he died in Geneva May 29, 1829.

Scientific Work.—The historical significance of Davy's scientific work lies chiefly in what he did to determine which substances ought to be considered as elements. Lavoisier had taken the most important step toward an answer to this question but he had left it in a rather unsatisfactory state. His list of elements (see page 56) was partly incomplete and partly dependent upon certain assumptions which we now recognize as inspirations of genius but which nevertheless still lacked experimental foundation. He included lime and alumina for example among his elements because, while probably oxides, they had not yet been decomposed, and he was so certain that the alkalies were oxides that he dropped them from the list. Silica he classified as an earth, and the acids of salt, borax and fluorspar he also considered oxides, and put in the list the *radicals* of those acids. It was the great service of Davy that partly by his own work and partly by that of others whom his researches inspired, most of the questions involved here were definitely settled.

Studies in Electricity.—In 1801 the scientific world was full of interest in voltaic electricity, and Davy, now that he was in charge of a well-equipped laboratory, threw himself with energy into the study of these phenomena. His early experiments were in electrolysis, and especially in the study of the various combinations of substances which would yield currents. One of the most important results was the discovery that active cells could be constructed which contained only one metal and two liquids. This work, however, was soon overshadowed by that of which he began to give an account in the Bakerian lecture of 1806. This investigation began with the consideration of a comparatively unimportant problem which, however, gained dignity by the scrupulous care and experimental skill which Davy employed in its solution.

Isolation of the Alkali Metals.—Nicholson and Carlisle, as well as others who had studied the electrolysis of water, had noticed the formation of acid and alkali at the poles, and concluded that the decomposition was by no means as simple a phenomenon as it seemed. Davy, however, showed that pure water is a much less common substance than is commonly supposed, and that samples previously electrolyzed had contained impurities derived either from the containing vessel or from the atmosphere. He then showed that pure water electrolyzed in vessels of gold yields only oxygen and hydrogen in chemically equivalent quantities. Davy had already obtained acids and bases by the electrolysis of salts, and he now tried the new force upon substances not hitherto decomposed. He caused to be constructed the most powerful battery then in existence, and with it attempted to decompose the alkalies. He first used strong solutions of potash, and then fusions of the dry alkali without results, but the following extract from the Bakerian lecture of 1807 describes his final success:

"A small piece of pure potash, which had been exposed for a few seconds to the atmosphere, so as to give conducting power to the surface, was placed upon an insulated disc of platina, connected with the negative side of the battery of the power of 250 of 6 and 4,[1] in a state of intense activity; and a platina wire, communicating with the positive side, was brought in contact with the upper surface of the alkali. The whole apparatus was in the open atmosphere.

[1] This means a voltaic battery of 250 pairs of plates each 6 by 4 inches in size.

"Under these circumstances a vivid action was soon observed to take place. The potash began to fuse at both its points of electrization. There was a violent effervescence at the upper surface; at the lower or negative surface, there was no liberation of elastic fluid; but small globules having a high metallic lustre, and being exactly similar in visible characters to quicksilver, appeared, some of which burnt with explosion and bright flame, as soon as they were formed, and others remained, and were merely tarnished, and finally covered by a white film which formed on their surfaces."

The isolation of potassium above described was soon followed by that of sodium, and Davy made an extensive study of the properties and chemical relations of each. The importance of the discovery and the surprising properties of the new metals aroused the greatest interest and Davy found himself world-famous almost in a day. What he had done also encouraged others to work along similar lines. Gay-Lussac and Thénard soon found that they could obtain sodium and potassium in still better yield by reduction with metallic iron, and hereafter a healthy rivalry sprang up between Davy and the French chemists which was on the whole for the benefit of all concerned.

Davy, like Lavoisier, regarded the caustic alkalies as oxides and the similarity of these substances to ammonia led him to attempt the decomposition of the latter. Here an unfortunate blunder made him state that the dry gas contained 7 or 8 per cent. of oxygen. This statement stood among others of the highest accuracy and had some unfortunate theoretical consequences, for many chemists were misled, and Davy's results for a time displaced those of Berthollet who had correctly determined the composition of the gas some years before. Davy also made a study of ammonium amalgam, and finding that it decomposed readily into ammonia and hydrogen he raised the question whether free sodium and potassium might not also contain hydrogen, a conclusion for which their combustibility seemed to speak. Not only Davy but also Gay-Lussac and Thénard considered this question seriously, but the latter chemists finally furnished the most definite proof that it could not be true. They burned potassium in dry oxygen, and found that no water was formed in the process. If the potassium contained any hydrogen it must therefore still be present in the peroxide. The latter, however, is readily decomposed by carbon dioxide into oxygen and potassium carbonate, in neither of which can any hydrogen be found.

Davy next accomplished the decomposition of the alkaline earths but this proved more difficult. Acting upon a suggestion of Berzelius, he mixed the earths with mercuric oxide and subjected the mixtures to the action of his battery. This yielded amalgams from which the metals could be prepared with some difficulty. Barium, strontium, calcium and magnesium were thus added to the list of elements known in the free state.

Davy now tested the effects of the current upon boric acid and silica, but without success. Nevertheless the former was soon decomposed by Gay-Lussac and Thénard who fused it with metallic potassium, and silicon was isolated by Berzelius who heated silica with iron and carbon. Alumina was not decomposed until 1827, but no one any longer doubted that it was an oxide. This left of the *radicals* of Lavoisier's list only those of muriatic and hydrofluoric acids. These, however, we must not confuse with the elements chlorine and fluorine. Instead they were hypothetical entities which had been introduced into chemical theory by Lavoisier's erroneous conception concerning the nature of acids. Since this conception was long dominant and destined to have unfortunate consequences for many years to come, we must divest ourselves of some modern ideas which now seem axiomatic and learn to think in its terms.

Lavoisier's Theory of Acids.—Rouelle had fixed the idea of a salt as an *addition product* of acid and base (page 32) and Lavoisier had incorporated this idea in his system. The term 'acid' was at this time universally applied to the substance which we call the anhydride, while what we call the acid was considered as the acid plus a certain quantity of more or less adventitious water—akin to water of crystallization. If Lavoisier had used our symbols he would therefore have formulated sulphuric acid as SO_3 and the reagent which we now call by that name SO_3,HO (Dalton's atomic weights). When this reacts with lime, calcium sulphate and water are produced, the latter coming entirely from the acid in the sense of the following equation:

$$CaO + SO_3,HO = CaO,SO_3 + HO$$

Here both acid and base are oxides, and Lavoisier doubtless believed that this was always the case, but while several of the bases had not been decomposed, there was no question that carbonic, sulphuric and phosphoric acids were oxides. He therefore believed that oxygen was the essential constituent in every acid, and gave the element its name on that account. It now remains for us to inquire how muriatic acid, for example, could be fitted into such a system. This was done rather ingeniously as follows:

The Muriaticum Theory.—Since all acids contain oxygen, muriatic can be no exception, and since the dry gas acts upon dry oxides like lime to form water it must also contain that compound, just as sulphuric acid does. The real anhydrous acid must therefore be the oxide of a radical as yet unknown but commonly called *murium* or *muriaticum*. If we designate this by X, then anhydrous muriatic acid will be XO, and the well-known gas HO,XO. What then is chlorine? Scheele had discovered this substance by treating muriatic acid with the black oxide of manganese and had naturally given to the product the name of "dephlogisticated marine acid." After the discovery of oxygen Berthollet changed this to "oxidized muriatic acid" or "oxymuriatic acid" which it had since retained. Everyone believed that it contained oxygen and this was supported by the oxidizing action of bleaching powder, the evolution of oxygen gas from chlorine water, and a number of other reactions which had really not been carefully studied. It will perhaps help us to acquire the point of view of the times if we formulate some of the familiar reactions of chlorine and muriatic acid both in the terms of the *muriaticum* theory and also in our modern formulæ:

(1) $NaO + HO,XO = HO + NaO,XO$; $Na_2O + 2HCl = 2NaCl + H_2O$.
(2) $HO,XO + O = HO + XO_2$; $2HCl + O = H_2O + Cl_2$.
(3) $H + XO_2 = HO,XO$; $H_2 + Cl_2 = 2HCl$.
(4) $Na + XO_2 = NaO,XO$; $2Na + Cl_2 = 2NaCl$.

It will be seen that the *muriaticum* theory 'explained' these reactions satisfactorily from the quantitative as well as the qualitative side, and in the Bakerian lecture of 1808 we find Davy entirely in accord with this view, though evidently rather disappointed that his attempts to prepare the anhydrous acid had been so unsuccessful. When the ordinary acid was treated with a dry base it lost its water readily enough:

$$NaO + HO,XO = HO + NaO,XO$$

but no water could be extracted from the gas by dehydrating agents, so he next attempted to obtain the anhydrous acid by decomposing a dry chloride with another acid (YO), in the sense of the equation:

$$NaO,XO + YO = NaO,YO + XO$$

Under no circumstances, however, could this reaction be made to take the indicated course. He heated numerous muriates with anhydrous boric acid and with dry silica but could get no trace of decomposition till moisture was admitted to the vessel. Then indeed muriatic acid was evolved but it was the familiar gas which he believed to contain water.

The Elementary Nature of Chlorine.—If muriatic acid contains oxygen then oxymuriatic acid (chlorine) must contain still more, but all Davy's attempts to obtain oxygen from it were equally fruitless. It might be supposed that if oxymuriatic acid contained oxygen phosphorus would remove it. Davy writes on this point as follows:

"I have described, on a former occasion, the nature of the operation of phosphorus on oxymuriatic acid, and I have stated that two compounds, one fluid, and the other solid, are formed in the process of combustion, of which the first, on the generally received theory of the nature of oxymuriatic acid, must be considered as a compound of muriatic acid and phosphorous acid, and the other of muriatic acid and phosphoric acid.[1] It occurred to me that if the acids of phosphorus really existed in these combinations, it would not be difficult to obtain them and thus to gain proofs of the existence of oxygen in oxymuriatic acid.

"I made a considerable quantity of the solid compound of oxymuriatic acid and phosphorus by combustion, and saturated it with ammonia, by heating it in a proper receiver filled with ammoniacal gas, on which it acted with great energy, producing much heat; and they formed a white opaque powder. Supposing that this substance was composed of the dry muriates and phosphates of ammonia; as muriate of ammonia is very volatile, and as ammonia is driven off from phosphoric acid, by

[1] $P_2 + 3XO_2 = P_2O_3, 3XO$
$P_2 + 5XO_2 = P_2O_5, 5XO$

a heat below redness, I conceived that by igniting the product obtained, I should procure phosphoric acid; I therefore introduced some of the powder into a tube of green glass, and heated it to redness, out of contact of air by a spirit lamp; but found, to my great surprise, that it was not at all volatile nor decomposable at this degree of heat, and it gave off no gaseous matter. * * * I contented myself by ascertaining that no substance known to contain oxygen could be procured from oxymuriatic acid, in this mode of operation."

The above experiment represents only one of many convincing ones, which showed that no oxygen could ever be obtained from chlorine or dry muriatic acid, and that when it appeared in reactions involving one of these substances its presence could always be accounted for in some other way. Davy accordingly held that it was simpler to assume that "oxymuriatic acid" is an element to which he now gave the name *chlorine* on account of its color. It is worth while to emphasize the fact that Davy did not prove that chlorine was not a compound, nor is such proof possible. Chlorine may still contain oxygen (or titanium for that matter) but so long as no other element can be obtained from it, its compound nature is a gratuitous assumption. Numerous chemists who admired Lavoisier and valued a logical system more than experimental evidence, preferred for some time longer to make this assumption, and Gay-Lussac and Thénard were particularly hard to convince. Indeed they supported the *muriaticum* theory with some very ingenious experiments. As fate would have it, however, they soon themselves furnished the most satisfactory evidence against the theory. In 1813 Gay-Lussac published a famous paper upon iodine, then recently discovered by Courtois, and in the following year another paper on cyanogen which is equally noteworthy. These investigations involved a study of hydriodic and hydrocyanic acids; and the important analogies which connect these with hydrochloric, together with the certainty that there is no oxygen in hydrocyanic acid soon satisfied all that there was no oxygen in any of them.

The Hydrogen Theory of Acids.—Davy drew the logical conclusion from these results, and was inclined to proclaim hydrogen and not oxygen as the essential component of acids. Unfortunately the influence of Lavoisier's name and the force of tradition prevented the general adoption of this idea. Instead scientists accepted the compromise suggested by Gay-Lussac which involved the creation of a new class of compounds called *hydracides* to include these acids which contained no oxygen. Hydrochloric, hydriodic and hydrocyanic acids obtained their present names at that time.

The above constitute Davy's most important investigations from the historical point of view. On the practical side, he invented the safety lamp for miners in 1817, and in 1825 he suggested that the copper with which the bottoms of ships were then universally sheathed might be protected from corrosion by the addition of comparatively small pieces of zinc. The corrosion was diminished, but the accumulation of marine growths upon the plates was so much accelerated that the idea proved impracticable.

Theory of Chemical Affinity.—Toward Dalton's atomic theory Davy maintained a sceptical attitude akin to that of Wollaston. His experiments with electricity, however, led him to a theory of chemical affinity which has interest as a sign of the times. Davy had at first been a supporter of the "chemical" view, but the repetition of some of Volta's experiments with the electrometer not only converted him to the contact theory, but led him to develop it into a general theory of chemical combination. He found, for example, that copper is always positive toward sulphur, and that this difference in polarity becomes accentuated with rise of temperature. Finally, the elements unite to form copper sulphide with an evolution of heat which Davy ascribed to the neutralization of the electric charges. Till this time chemists had compared chemical affinity to gravitation, and some had gone so far as to suggest the identity of the two forces. Now, however, it was natural that Davy and his contemporaries should identify chemical affinity with electricity. Davy held that when the atoms of two substances are brought into proximity they assume opposite electrical charges and finally unite as the charges are neutralized. This gives a consistent explanation of electrolytic decompositions by the current. The constituent elements receive from the poles of the battery the electric charges which they possessed before combination, and they can henceforward again exist in the free state. This idea was never developed into a complete system by Davy, but the same general idea lay at the foundation of the theories advanced later by Berzelius.

Davy and Faraday.—It has often been said that Davy's greatest discovery was Michael Faraday. Faraday, who was born in Newington, Surrey, in 1791, was of very humble parentage and began life as a bookbinder's apprentice in 1814. He not only became an expert in this handicraft but by reading the books as well as binding them[1] he gradually acquired a considerable knowledge of many subjects which interested him, particularly things which had to do with natural science. The story of his meeting with Davy is told in the following letter.

To J. A. Paris, M. D.
Royal Institution, December 23, 1829.

My dear Sir:—You asked me to give you an account of my first introduction to Sir H. Davy, which I am very happy to do, as I think the circumstances will bear testimony to the goodness of his heart.

When I was a bookseller's apprentice I was very fond of experiment and very averse to trade. It happened that a gentleman, a member of the Royal Institution, took me to hear some of Sir H. Davy's last lectures in Albemarle Street. I took notes, and afterwards wrote them out more fairly in a quarto volume.

My desire to escape from trade, which I thought vicious and selfish, and to enter into the service of Science, which I imagined made its pursuers amiable and liberal, induced me at last to take the bold and simple step of writing to Sir H. Davy, expressing my wishes, and a hope that if an opportunity came in his way he would favor my views; at the same time I sent the notes I had taken of his lectures.

The answer, which makes all the point of my communication, I send you in the original, requesting you to take great care of it, and to let me have it back, for you may imagine how much I value it.

You will observe that this took place at the end of the year 1812; and early in 1813 he requested to see me, and told me of the situation of assistant in the laboratory of the Royal Institution, then just vacant.

At the same time he thus gratified my desires as to scientific employment, he still advised me not to give up the prospects I had before me, telling me that Science was a hard mistress, and in a pecuniary point of view but poorly rewarding those who devoted themselves to her service. He smiled at my notion of the superior moral feelings of scientific men, and said he would leave me to the experience of a few years to set me right on that matter.

Finally, through his good efforts, I went to the Royal Institution early in March of 1813, as assistant in the laboratory; and in October

[1] The phrase is Ostwald's.

of the same year went with him abroad, as his assistant in experiments and in writing. I returned with him in April, 1815, resumed my station in the Royal Institution, and have, as you know, ever since remained there.

I am, dear Sir, ever truly yours,
M. Faraday.

When Faraday accepted this position it commanded a salary of 25 shillings a week and the use of two small rooms in the upper story. The incumbent could hardly be said to have deserved more, for he had never enjoyed anything but the most elementary schooling, and had had no regular training in science. It is splendid evidence of his genius that within two or three years he was making discoveries in both chemistry and physics which rivalled in quality those of his master. As we all know, his later studies on the relations of electricity and magnetism stand at the foundation of modern electrical engineering. He did pioneer work in the liquefaction of gases, and among his other discoveries which interest chemists are that of magnetic optical rotation, the isolation of benzene and of the two sulphonic acids of naphthalene. It may be appropriate to add that to Faraday's friends his scientific attainments were surpassed by the peculiar charm of his personality. We shall have occasion to discuss some of his further contributions to chemistry later on. He succeeded Davy at the Royal Institution in 1825 and spent the remainder of his active life there. He died in 1867.

Literature

The Works of Humphry Davy (in nine volumes) were published by his brother John Davy in 1834. The first volume is biographical. T. E. Thorpe has also published a life entitled *Humphry Davy, Poet and Philosopher*, London, 1896. Selections from his work on chlorine have been reprinted by the Alembic Club, No. 10. The latter paper should also be read in connection with No. 13 of the same series, which gives an idea of the earlier views as to the nature of chlorine, as expressed by Scheele, Berthollet, Guyton de Morveau, Gay-Lussac, and Thénard. Davy's work leading to the discovery of the alkali metals may be found in *Alembic Club Reprint* No. 6.

Among the better books on Faraday may be mentioned the *Life and Letters* by Bence Jones, London, 1870, and *Faraday as a Discoverer* by John Tyndall, London, 1870.

Ostwald devotes a chapter each to Davy and Faraday in his *Grosse Männer*, Leipzig, 1909, and Thorpe's *Essays* contain a chapter on Faraday.

Jöns Jakob Berzelius
1779–1848

(Facing page 100)

Eilhardt Mitscherlich
1794–1863

CHAPTER XI

:RZELIUS, THE ORGANIZER OF THE SCIENCE

s Jakob Berzelius was born in Väfersunda, Sweden, August '79. His father, who was a teacher in Linköping, died in and the mother's second marriage, soon followed by her death, left the young Berzelius to grow up in the home ıtives where his slender means made him a not altogether me guest. These experiences did much to embitter his years, and matters were hardly improved at the *gymnasium* he distinguished himself, indeed, by love of natural history ›ok little interest in the classics, and was unwilling to make ttempt to win the favor of those whose specialties he dis- On his departure the authorities handed him a certificate g that he "justified only doubtful hopes"—an amusing ɩentary on the perspicacity of teachers.

'zelius next studied medicine at Upsala but found the :es rather poorly represented there, and the same fatality had pursued him at the *gymnasium* involved him in tunate misunderstandings with the professors. These nostly to do with formalities but they prevented cordial ›ns. It may encourage those dissatisfied with academic ards to know that when Berzelius came to be examined in stry the professor in charge stated that he deserved to fail, :xpressed willingness to overlook the candidate's short- ıgs if he could give a better account of himself in physics ch he fortunately did.

'zelius meantime went on with chemical investigations d out partly in the laboratory and partly in his own rooms. 02, all examinations being over, he went to Stockholm and up hospital work devoting his spare hours to chemistry. vork done soon attracted the attention of men connected the college of medicine, and Berzelius was gradually drawn :loser touch with that institution, first as "adjunct" and finally as professor. As he grew famous, honors, medals, titles and emoluments were showered upon him in a profusion which must have made some amends for the hardships of his early life. He died August 7, 1848.

Work on the Combining Weights.—Among the earliest investigations of Berzelius were some upon electricity which we shall take up later, but about 1810 there began to appear from his pen a series of articles entitled: *An Attempt to Determine the Definite and Simple Proportions in which the Constituents of the Inorganic World are Combined with Each Other.* These articles continued till about 1818 and represented work begun as early as 1807. In one sense it was a topic with which Berzelius was actively concerned until his death. He had been much impressed with the ideas of Dalton and Richter, but he had seen more clearly, perhaps, than any one else, that they could never become the basis of a system until they were supported by an experimental foundation unattainable by men so inferior in analytical skill. Berzelius therefore set himself no less a task than to determine with the utmost possible accuracy the combining weights of the elements; and, within a little more than ten years, he accomplished this for about fifty of them by the preparation, purification and analysis of no less than two thousand of their compounds with his own hands. This was done at a time when quantitative analysis as we now understand the term hardly existed, and Berzelius was obliged in the majority of cases to laboriously work out his methods as he went along. We have also to remember that in his time the reagents had for the most part to be themselves prepared or extensively purified, and that the laboratory facilities of Berzelius boasted little beyond those afforded by an ordinary kitchen. The results speak for themselves. Below stands a selected list in which some of the atomic weights published by Berzelius in 1826 are com-

	Berzelius, 1826	International commission, 1917
Lead	207.12	207.20
Chlorine	35.41	35.46
Potassium	39.19	39.10
Sulphur	32.18	32.06
Silver	108.12	107.88
Nitrogen	14.05	14.01

pared[1] with those of the international commission for 1917. When we have admitted that in the case of many of the less common elements the divergence was much larger, and that the surprising agreement was sometimes due to a fortunate balancing of errors not appreciated by Berzelius, the achievement still stands as a remarkable monument to his genius. Henceforward quantitative analysis was destined to stand essentially upon its modern footing and no theory unable to square with its standards could any longer receive consideration.

Literary Activity.—Nor was this by any means the only occupation of Berzelius during these years. In addition to his activity as teacher he published no less than thirty papers dealing with other chemical problems and he developed a complete system of chemical philosophy which, in logical unity and comprehensiveness of scope, surpassed anything hitherto known. This he communicated to the world partly through the pages of his famous *Jahresbericht* which he founded in 1810, and continued to edit until his death, and partly by the great text-book which he began in 1808 and which passed through five editions. The minute first-hand knowledge of his facts which rings through every line of these writings, the hitherto unequalled accuracy of his results, and his unquestioned integrity gave to Berzelius a position altogether unique, so that in 1820 he ranked as a kind of law-giver whose mere opinions often counted more in the public mind than facts and figures carefully ascertained by others. Such a position of acknowledged superiority doubtless contributed to make him autocratic and intolerant in his later years, but for us the important consequence is that his views had the utmost influence in every department of the science. We must therefore consider them in detail.

Chemical Problems at the Beginning of the Nineteenth Century.—The moot points in chemical theory in the beginning of the nineteenth century we already know. They can be stated in questions as follows: How are chemical compounds to be formulated? Can atomic weights (as distinguished from combining

[1] Berzelius's standard of reference was O = 100. The results in this table are calculated to O = 16 and made comparable with our own by the use of modern formula-weights which sometimes differ from those of Berzelius for reasons which we shall appreciate more clearly later on.

weights) be determined? What is the essential constituent of acids? What is the source of the galvanic current, and what is the mechanism of its action in electrolysis? What is chemical affinity and is any quantitative measure thereof possible? On all of these subjects Berzelius had decided views supported by an immense amount of experimental data.

He did much to facilitate all chemical discussion and calculation by inventing our modern alphabetical chemical symbols which were entirely original with him, and by developing a system of chemical nomenclature which was essentially an adaptation of that of Lavoisier to the Germanic languages. Indeed, we shall best understand the spirit which animated the work of Berzelius if we consider him as an admirer and disciple of Lavoisier who consciously made it his life-work to extend and complete the chemical system which the great Frenchman had left unfinished.

The Atomic Weights.—This manifested itself everywhere, even in the question of the atomic weights. Here Berzelius saw at once the hopelessness of Dalton's assumption that the best-known compound must have the simplest formula, and he looked about for other criteria. He was satisfied that he had found one in Gay-Lussac's law of combining gas volumes and he adopted as his principal standard the assumption that the atomic weights of *elementary* gases are proportional to the weights of equal volumes. This, however, at that time could not carry him far. Dalton had already pointed out that equal volumes of *compound* gases could not contain equal atoms, and while the discrepancy involved here might have been obviated by an intelligent application of Avogadro's hypothesis, Berzelius was by temperament opposed to such a notion, and declined to consider an explanation which involved a contradiction in terms so flagrant as fractions of atoms. He was therefore limited to the elementary gases and these were so few in number that they might have appeared as likely to represent the exception as the rule. For the majority of the elements some other auxiliary standard was necessary and Berzelius found this in the varying quantities of oxygen with which the various elements could combine. This was partly because oxygen combines so freely with almost all other elements, but quite as truly because of the exaggerated opinion of the importance of oxygen which he had inherited from Lavoisier (see page 58). Berzelius once wrote, "Oxygen is the center about which all chemistry revolves," and similar expressions are common in his writings. His work upon the combining weights and the analysis of salts had led him to formulate the rule that when an acid unites with a base the ratio of the oxygen in the acid to that in the base is a simple whole number.

The method by which Berzelius worked this out in a specific case may be of interest. He prepared lead sulphate by oxidation of the sulphide with nitric acid, and showed qualitatively that no lead or sulphuric acid was left in excess in the supernatant liquid. It followed that the ratio of lead to sulphur must be the same in the sulphate as in the sulphide. Lead sulphate, however, was regarded as a binary compound of lead oxide and sulphuric acid (what we now term the anhydride). Since now the amount of oxygen combined with a given amount of lead in the oxide is already known, the balance of it—or three times that quantity—must in the sulphate be combined with the sulphur, and the number of atoms of oxygen in sulphuric acid must be three or some multiple of three. In the absence of any evidence requiring a larger number, Berzelius's assumption would be that it was exactly three, and we could now employ these data to determine the atomic weight of sulphur, for since the atomic weight of oxygen is fixed by its density at 16, the formula SO_3 for sulphuric acid fixes the atomic weight of sulphur at 32. Turning now to the basic component in lead sulphate we might further argue that *if* the atomic weight of lead is that quantity which unites with 16 units of oxygen then the same is likely to hold true for metals like barium and calcium. The validity of the fundamental assumption as well as that of the suggested analogy would, however, in this case have to be supported by other evidence of a similar kind. It is easy to see that the reasoning of Berzelius is superior to that of Dalton, but it is also clear that he really makes use of the latter's principle of maximum simplicity, and that arbitrary assumptions have by no means been excluded. When, in such a case, the sum total of assumptions leads to a system free from contradictions we can attach a high degree of probability to the results. The system of Berzelius never quite reached that stage, and he himself was so conscious of its imperfections that we find him modifying his figures repeatedly as long as he lived.

Law of Dulong and Petit.—In 1819 two papers appeared which threw fresh light upon the subject because they suggested new criteria. The first of these was by Dulong and Petit, and called attention to the remarkable relationship which exists between the atomic weights and the specific heats. The authors had been determining the latter constant for a large number of substances, and were impressed with the fact that in the case of most of the solid elements, when the specific heats were multiplied by the atomic weights then current, the result was a constant. They also expressed the results in the form that "the atoms of all the elements have the same heat capacity," that is to say if a certain quantity of heat will warm 63 grams of copper one degree it will do the same for 56 grams of iron. Where the constant product above referred to was not obtained, it could be in most cases reached by multiplying or dividing the atomic weight employed by a simple factor. Dulong and Petit made bold to take this step, feeling that they had here a new standard resting upon a measurable physical constant, and free from the assumptions involved in the reasoning of Berzelius and Dalton.

We now know that the law is not sufficiently infallible to justify such disregard of all other considerations. Some of the determinations of Dulong and Petit were inaccurate, in other cases the specific heat varies widely with the temperature, involving an embarrassment in the selection of comparable conditions, and finally, for reasons which are even now not completely understood, the law seems unreliable in the case of the elements of lowest atomic weight. Berzelius's own attitude in the matter was cautious and conservative. He threw some doubt upon the accuracy of the determinations, but later was persuaded to divide the atomic weight of silver which he had previously adopted by two. He declined, however, to modify the atomic weight of carbon as the theory of Dulong and Petit seemed to demand, because that would have made the oxides CO_2 and CO_4 respectively, which chemical reasons did not permit him to accept. Time has amply justified his decision.

Isomorphism.—The other paper referred to was by Eilhard Mitscherlich (1794–1863), soon after a student of Berzelius and later professor in Berlin, where he succeeded Klaproth. He made some valuable studies of vapor densities and the simpler compounds of benzene, then a rare substance. His principal work, however, was in those departments of chemistry most allied to mineralogy and crystallography, of which the present work on isomorphism was a most auspicious beginning. Earlier

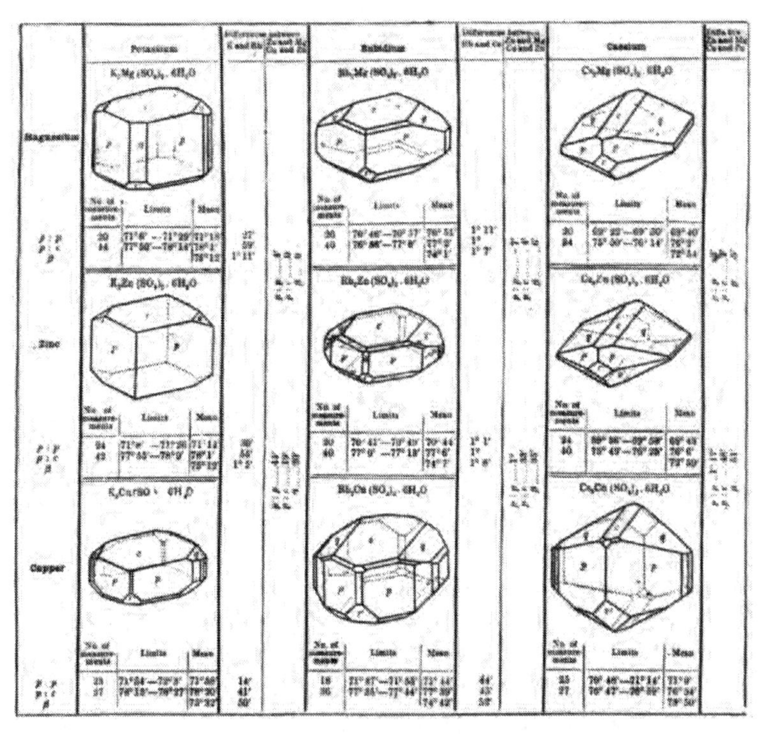

A Typical Group of Isomorphic Substances

Reproduced from Freund's "The Study of Chemical Composition" by the kind permission of the author and the Cambridge University Press.

investigators had sometimes noted marked similarity of crystalline form among different substances as well as the formation of certain mixed crystals. Mitscherlich, however, was at this time entirely ignorant of these observations, and in fact was just beginning his studies in crystallography. He also carried his observations much farther than others had done and refe-

back the now familiar phenomena of isomorphism to similarity of chemical composition. Working at first with the phosphates and arsenates he found that whatever metal was taken as a base there was always to be found the most perfect similarity in the properties of all the salts, not only in crystalline form but also in solubility and other properties, the similarity extending to the quantity of water of crystallization to be found in each. Examination of other series such as the alums and vitriols soon showed that this was no isolated case and Mitscherlich came to the conclusion that

"The same number of atoms combined in the same manner produce the same crystalline form. This form is independent of the nature of the atoms and is fixed only by their number and mode of combination."

It will be seen that so far as the choice of atomic weight is concerned isomorphism does not offer an entirely independent criterion like that of the specific heats, but rather an aid in reasoning by analogy which may be employed as follows: If we formulate ordinary alum as $KAl(SO_4)_2.12H_2O$ and the atomic weights of potassium and aluminium are considered as known quantities, then the atomic weights of sodium and iron are the parts of weight of those elements which unite with 64 parts of sulphur in ferric sodium alum, $FeNa(SO_4)_2,12H_2O$.

Berzelius gave rather more weight to such considerations than to those suggested by Dulong and Petit and he was induced thereby to modify his views concerning the atomic weight of chromium and some other elements. His reasons are best stated in his own words because they throw light upon his whole method of reasoning with regard to atomic weights.[1]

"It is known that the oxide of chromium contains three atoms of oxygen. Chromic acid for the same number of chromium atoms contains twice as much oxygen, which would be six atoms; but in its neutral salts chromic acid neutralizes an amount of base containing one-third as much oxygen as it contains itself, a relation found to hold in the case of all acids with three atoms of oxygen (*e.g.*, sulphuric acid and sulphates). In order to harmonize the multiple relation between the amount of oxygen in the oxide and in the acid, it is most probable that the acid contains three atoms of oxygen to one

[1] The quotation follows the translation of the passage by IDA FREUND in her admirable book, *The Study of Chemical Composition*.

atom of chromium, and the oxide three atoms of oxygen to two of chromium. Isomorphous with the oxide of chromium are those of manganese, iron and aluminium; these also we know to contain three atoms of oxygen, and consequently must represent them as containing two atoms of the radical. But if the ferric oxide consists of 2Fe + 3O, the ferrous oxide is Fe + O, and the whole series of oxides isomorphous with it contains one atom of the radical and one atom of oxygen. * * * Unfortunately in these matters the certainty of our knowledge is as yet at so low a level that all we can do is to follow along the lines of greatest probability."

The Dualistic System.—It is in his views on chemical composition and the nature of acids that Berzelius reveals most strongly the influence of Lavoisier. It is estimated that in his *Traité Élémentaire* the latter included approximately nine hundred substances, and of all this number there were only about thirty, aside from the elements, which could not be classified as acids, bases or salts. It is therefore entirely natural that Lavoisier should regard all chemical union as allied to salt formation. His views on the latter subject we understand already. Salts were binary addition products of acid and base, and these in their turn were binary addition products of metal and oxygen and of non-metal and oxygen respectively. Union occurred in pairs, and the system was distinctly dualistic.

Lavoisier regarded oxygen as the acid-forming element and yet made it just as necessary a component of bases. There is only an apparent contradiction here because the general tendency of oxygen was regarded as acidic though the other element might overcome this tendency, thus chromium with a certain amount of oxygen forms a base but with more oxygen an acid. Lavoisier too was prevented from carrying his generalizations to their logical conclusion by the fact that in his day many of the bases had not yet been proved to be oxides. Davy's discovery of the alkali metals removed this difficulty and left Lavoisier's system of chemical composition essentially complete, save for the troublesome case of hydrochloric acid among the acids and ammonia among the bases. We have already seen in the previous chapter the difficulty which Davy had encountered in convincing his contemporaries that chlorine did not contain oxygen.

Berzelius and the Chlorine Theory.—Berzelius was among those who fought most stubbornly in the defense of the older view. Even as late as 1815 he writes an article a hundred pages long in which he uses every argument to show that the old theory was still capable of explaining the facts, and urging all chemists to retain it in the interest of the unity of the science. He closes with the words:

"I demand of every chemical principle that it agree with the sum total of chemical theory and be capable of incorporation therewith. Otherwise I must reject it until such time as incontestable evidence in its favor makes it necessary to recast the entire system."

It is as necessary in science as in politics that there should be a conservative party, and it is always well that some one should speak in such terms as those just quoted, but the immense authority of Berzelius and the increasing obstinacy which characterized him in his later years did much to retard the acceptance of many new and helpful ideas. This intolerance of Berzelius has become so associated with most that is written about him that there is a real danger that those who know him only through tradition or his polemical writings will not appreciate the genial, tolerant and open-minded side of his character. As late as 1836 when the hydrogen theory of acids was again under discussion from another point of view we find him writing to Heinrich Rose:

"The experiments with anhydrous sulphuric acid and chlorides interest me greatly. They appear to lead to the long-expected conclusion that strong sulphuric acid is not SO_3 or even $SO_3 + H_2O$ but $SO_4 + H_2$, and that the same theory holds good for the haloid salts as for those of the oxygen acids."

and again two years later to Liebig on the same subject:

"I shall be entirely satisfied to place the new view beside the old as another means of explanation, and after all what are any of our views but means of explanation?"

It is only fair to say, however, that this apparent conversion to the hydrogen theory was only temporary and that he published nothing in its favor. So far as the earlier controversy was concerned, the work of Gay-Lussac upon the cyanides and iodides showed Berzelius that the old view was untenable, and by

1820 he gave up the contest and accepted the compromise suggested by Gay-Lussac and by Dulong according to which there could be two kinds of acids, the oxygen acids which formed salts directly by the addition of metallic oxides, and the halogen acids (or hydro acids) which formed salts by substitution of that hydrogen by a metal. Berzelius was loath to draw this distinction between substances so similar in all other properties, but he could in this way save his dualistic system, oxygen salts being still binary compounds of acid and base, while haloid salts were binary compounds of metal and halogen.

Supposed Oxygen Content of Ammonia.—The case of ammonia deserves a word of comment. Though markedly basic it differed from all other bases in that it contained no oxygen. When therefore in 1808 Davy thought he had discovered evidence that it did contain this element Berzelius grasped at the idea, and supported it for many years. Even the discovery by Henry and the younger Berthollet that ammonia could be decomposed quantitatively into nitrogen and hydrogen did not suffice to drive the idea from his mind. He next assumed that nitrogen was itself an oxide, and the atomic weight of *nitricum*, its hypothetical radical, appeared for a long time in his tables of atomic weights. The arguments of Berzelius were ingenious but we need not enumerate them here as they had no historical significance. He had surrendered the point completely by 1822, and was doubtless assisted to the change of view by a suggestion of Ampère to the effect that when ammonia formed salts it first added water yielding ammonium oxide, which could then form addition products with acids in a manner entirely orthodox.

Despite these minor concessions the dualistic system still remained a consistent whole. Inorganic substances could still be classified as elements, binary compounds of elements (like chlorides, sulphides and oxides), and finally as salts, which were either binary compounds of the elements or of the oxides (acids and bases). Salts could further combine with acids to form still other binary compounds which we call acid salts, with bases to form basic salts, or with other salts to form double salts.

During the life of Lavoisier this dualism which seemed to run through all inorganic nature lacked any adequate theoretical explanation. There seemed to be no good reason why other forms of combination should not occur, and various ternary compounds were recognized and classified as such. With the beginning of the nineteenth century, however, a new force had come upon the scene, and nearly all chemists now regarded electricity either as identical with chemical affinity or closely connected with it.

Electrical Explanation of Chemical Action.—Soon after leaving the university Berzelius, in collaboration with his friend Hisinger, had carried on a series of experiments upon the electrolysis of salt solutions with as large a battery as their means permitted them to construct, and had observed the separation of salts into acid and base which is now so familiar a phenomenon. These experiments made a great impression upon Berzelius and led him to set electricity as well as oxygen at the foundation of his chemical system. This was based on ideas not dissimilar to those of Davy, but with characteristic energy and passion for detail he developed them much further and won them a wider recognition. Davy had assumed that when the atoms of different elements approach each other they assume opposite electrical charges of greater or lesser intensity according to the nature of the substances concerned. Berzelius was even more mechanical in his conceptions, and assigned to every atom two poles like those of a magnet, upon one of which was concentrated positive and upon the other negative electricity. For the same atom, however, the quantity or intensity of the charge (Berzelius was not clear on this point) was by no means equal. Chlorine, for example, possessed an excess of negative, the alkalies of positive electricity. Oxygen was to Berzelius the "absolutely negative" element, and he placed this at one end of a series, in which potassium occupied the other extreme. Between stood the other elements, positive or negative according to their relative positions, hydrogen being at or near the neutral point. This was of course merely another form of the potential series, derived in this case from chemical considerations.

Electrical Explanation of Dualism.—The above was in beautiful harmony with the dualistic system already outlined and furnished a complete explanation for it. Sulphur was positive toward oxygen and therefore could unite with it to form sulphuric acid, which in its turn was by no means neutral, but rather, on account of its large oxygen content, decidedly negative. In the same way positive calcium united with the negative oxygen to form lime, in which the electrical character of the metal predominated, so that it was distinctly electropositive. Finally sulphuric acid and lime could themselves unite to form a salt, calcium sulphate CaO,SO_3, much more nearly neutral but not necessarily absolutely so, for a difference in charge had to be assumed for salts in order to account for the formation of double salts like alum.

One limitation the theory brought with it. Ternary compounds were no longer possible. Substances united with each other because they were positive or negative, hence union could only take place in pairs, and substances apparently ternary like potassium cyanide must be regarded as compounds of an element with a binary compound.

Berzelius sums the matter up in the following frequently quoted passage:

"If these electrochemical views are correct, it follows that every chemical compound depends entirely and alone upon the two opposite forces of positive and negative electricity; and therefore every compound substance consists of two parts united by the action of their electrochemical character since no third force exists. Hence it follows that every compound substance, whatever the number of its components, can be divided into two parts of which one is positively and the other negatively electric; for example sulphate of soda is not composed of sulphur, sodium and oxygen, but of sulphuric acid and soda, and these in their turn, can be separated into positive and negative components. In the same way alum cannot be considered as a compound directly of its elements, but is to be looked upon as the product of the reaction between sulphate of alumina as the negative element and sulphate of potash as the positive element."

The formula here suggested for alum, $K_2O,SO_3 + Al_2O_3,3SO_3 + 24H_2O$, is interesting to the modern reader as an extreme case of dualistic formulation. It will be observed that in all these cases the theory tacitly involves the continued independent existence of the oxide within the salt. In lead sulphate, for example, one-fourth of the oxygen is still thought of as combined with the lead and the rest with the sulphur.

Berzelius's Explanation of Electrolysis.—It will be seen that the theory of chemical composition just outlined is admirably adapted to explain all the facts then known concerning electrolysis. The current merely separates the salt into the positive and negative components of which it is composed and these appear primarily at the poles. The effect can, of course, be obscured in individual cases by secondary reactions. An acid according to Berzelius is not decomposed by the current. It simply increases the conductance of the water, which alone is decomposed into oxygen and hydrogen. A salt like potassium sulphate, on the other hand, decomposes into potassium oxide and sulphuric acid, both of which are hydrated by the water. The evolution of oxygen and hydrogen at the poles is, however, due solely to the simultaneous decomposition of the water. When a metallic salt like zinc sulphate is electrolyzed neither the acid nor the water is decomposed, but only the base, metallic zinc appearing at one pole and the oxygen at the other. Some contemporaries of Berzelius thought that zinc sulphate behaved primarily like potassium sulphate—the hydrogen formed by the decomposition of the water then reducing the zinc oxide originally formed at the negative pole. He himself, however, rejected this notion, because under ordinary conditions zinc slowly decomposes water with evolution of hydrogen. With reference to the origin of the current Berzelius like Davy at first supported the 'chemical' theory, and like him went over later to the other side. A brief account of his reasons can best be given when we come to consider his attitude toward Faraday's law.

The foregoing may serve as a rather inadequate outline of the dualistic electrochemical system of Berzelius as it stood practically complete in the years which immediately followed 1820. It was almost universally accepted, for, in spite of doubtful points here and there, it furnished a consistent, reasonably satisfactory explanation of practically every known reaction in the inorganic field. Though it was before long to meet vigorous attacks under which it ultimately succumbed, yet it constituted the most important single factor in chemical theory for many years, and fragments of it remain incorporated in our phraseology to this day. If Berzelius did not originate all its fundamental ideas it was nevertheless he who practically single-handed

FRIEDRICH WÖHLER
1800-1882

e it unity and vitality, and as we have seen he enjoyed a 'esponding prestige.

Vöhler's Reminiscence.—We have a pleasant personal remi-:ence of Berzelius and his surroundings at just this time when vas at the height of his fame, from the pen of Friedrich Wöhler. some years it had been the habit of Berzelius to invite to his se by ones and twos certain young chemists of thorough train- and great promise, and permit them to spend a year or more iis laboratory. This was, of course, a wonderful educational ortunity and of those who had the advantage of it few failed achieve marked distinction in later years. Wöhler was in ckholm during the winter of 1823–24 and he has given an ount of his experiences in an article entitled *Jugenderin-ingen eines Chemikers.*[1] It should be quoted entire, but we e room for only the following fragments:

With beating heart I stood at Berzelius's door and rang the bell. eatly dressed, stately man of fresh appearance opened. It was :elius himself. He welcomed me in a most friendly way, said that ad been expecting me for a long time, and talked about my journey, ourse all in the German language, in which he was as proficient n French and English. When he took me into his laboratory I as if in a dream, doubting if it was a reality to see myself in these sic rooms, and so at the goal of my wishes.

On the following morning I began work. I obtained for my special a platinum crucible, a balance with weights, a wash-bottle, and re all a blow-pipe upon the use of which Berzelius laid great stress. ny own expense I had also to provide alcohol for the lamps and or the blast-lamp; the ordinary reagents and utensils were had in mon; but ferrocyanide of potassium, for example, was not to be had tockholm so I had to order it from Lübec. I was at that time the one in the laboratory; before me Mitscherlich and H. and G. Rose been there, and after me came Magnus. The laboratory consisted wo ordinary rooms with the very simplest arrangements: there : neither furnaces nor hoods, neither water system nor gas. In one ie rooms there were two ordinary long work-tables of pine wood, ie of them Berzelius had his place, at the other I had mine. Against walls stood some closets with the reagents, in the middle the mercury gh and the blast-lamp table, the latter under a flue leading into chimney of the stove. Beside these was the sink, consisting of a

[1] *Berichte der Deutschen Chemischen Gesellschaft,* vol. 8, p. 838 (1875).

stone water-holder with a stop-cock and a pot standing under it where each day the severe Anna the cook had to wash the dishes. In the other room were the balances and some presses with instruments and utensils; nearby still another little workshop with a lathe. In the kitchen close by, in which Anna prepared the food, stood a small heating furnace seldom used and the constantly heated sand-bath.

* * * * * * * *

"In the investigation of cyanic acid which I took up again Berzelius interested himself very much. To his great satisfaction he showed me what he had said in his *Jahresbericht* about my earlier experiments with this acid, and expressed the opinion that the existence of the same had contributed much to the greater probability of the chlorine theory. I was much surprised to hear him now speaking of chlorine instead of oxidized muriatic acid as, up to this time, he had been a firm defender of the old opinion. Once when Anna was cleaning a dish, she remarked that it smelled strongly of oxidized muriatic acid, Berzelius said, "Listen, Anna, you must not say oxidized muriatic acid any more. Say chlorine, it is better."

* * * * * * * *

"By repeated operations we had such quantities of potassium as had never before been produced. For the analyses at that time, we prepared pure caustic potash by burning potassium on water. Berzelius was as a rule cheerful, and during the work he used to relate all sorts of fun, and could laugh right heartily over a good story. If he was in bad humor and had red eyes, one knew that he had an attack of his periodic nervous headache; he would then shut himself up for days together, ate nothing and saw no one. A new observation always gave him great pleasure and with beaming eyes he would then call to me, "Well, Doctor, I have found something interesting."

"Sometimes Berzelius kept me with him in the evening, when the talk was on his journeys in France and England, on Gay-Lussac, Thénard, Dulong, Wollaston, H. Davy and other distinguished men of science of that period, upon whose shoulders we of a later generation now stand, all of whom he knew personally and whose individuality he well understood how to characterize. Chief in his esteem and veneration were Gay-Lussac and Humphry Davy; of the latter he always spoke with the greatest admiration of his genius. He corresponded with them all and preserved their letters. With pleasure I took advantage of his permission to read them, and later too he gave me his interesting journals of his travels to read, which contained a full account of his visits to Paris and London."

The article closes with an interesting glimpse of Davy who just ıen happened to be enjoying a vacation trip in Sweden.

Literature

A German translation of BERZELIUS's *Selbstbiographische Aufzeichnungen* ›pears as No. 7 in KAHLBAUM's attractive series entitled *Monographien ıs der Geschichte der Chemie*, Leipzig, 1903. An account of Berzelius's e down to 1821 by SÖDERBAUM constitutes No. 3 of the same series.

The reminiscences by WÖHLER are in the *Berichte der Deutschen Chemischen* *sellschaft*, vol. 8, p. 838 (1875). Berzelius's early work on atomic weights ›wn to 1812 is in No. 35 of the *Klassiker*.

From this point on the thorough student of chemical history must depend ›re and more upon the journals, of which the most important during the st part of the nineteenth century were BERZELIUS's *Jahresbericht*, the *ınales de la Chimie et de la Physique* which contain most of the work of ench chemists, and LIEBIG's *Annalen der Chemie und Pharmacie* containing ›st of that published by Germans.

CHAPTER XII

DUALISM IN ORGANIC CHEMISTRY—WÖHLER, LIEBIG AND DUMAS

Before considering the fate of Berzelius's dualistic system after the attempt was made to apply its principles to organic chemistry, it will be necessary to introduce three younger chemists destined to take an active part in dealing with the problems which now pressed for solution. These were Wöhler, Liebig and Dumas.

Wöhler.—Friedrich Wöhler was born in Eschersheim near Frankfurt in 1800. He studied medicine at Marburg and Heidelberg where he came under the influence of Leopold Gmelin (1788–1853) who aroused his interest in chemistry and sent him with warm recommendations to Berzelius. Of his life at Stockholm we have had some account in the last chapter. He returned to Germany in 1824 as teacher in the *Gewerbeschule* at Berlin where he remained until 1831. In that year he became professor in a similar institution recently founded at Cassel. In 1836 he accepted the professorship of chemistry at Göttingen where he remained till his death in 1882.

Wöhler's scientific work covered a wide field and we shall have frequent occasion to refer to it. Among those investigations which have less historical significance may be mentioned his discovery of aluminium in 1827 and his work upon boron, silicon and titanium. In the organic field he added much to our knowledge of the cyanates and other substances of this class, while he practically laid the foundation for all subsequent work in the study of uric acid. He was highly distinguished as a teacher and drew to Göttingen many students from other countries, especially from America.

Liebig.—Justus Liebig was born in Darmstadt in 1803. His father did a small business in oils, colors and the more common chemicals, many of which he prepared himself. In such a laboratory and the work-shops of artisans in the vicinity Liebig

Justus Liebig
1803–1873

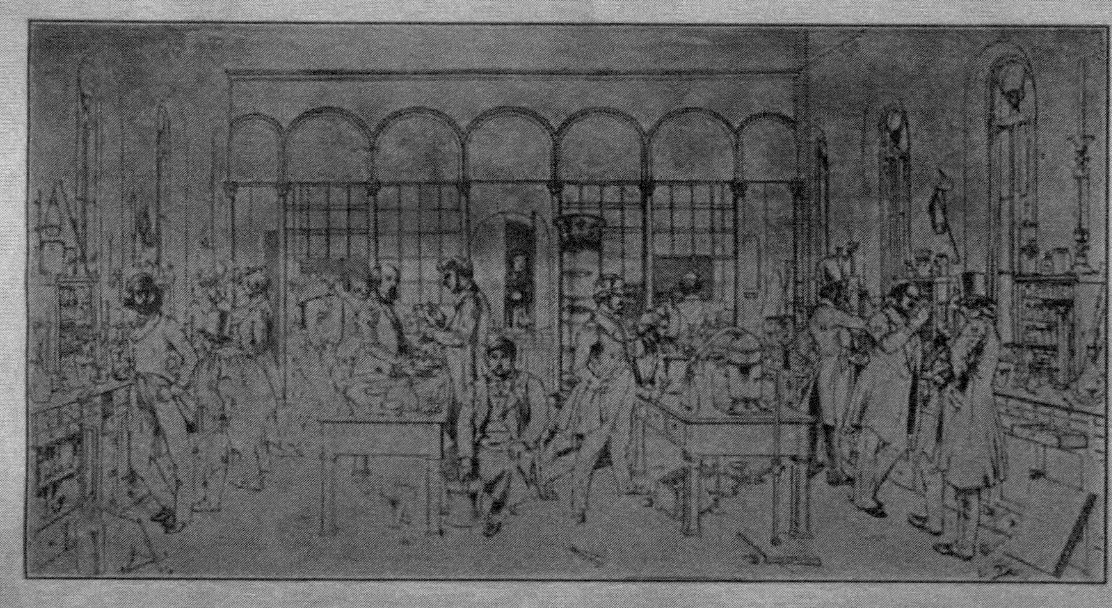

INTERIOR OF LIEBIG'S LABORATORY AT GIESSEN

From the Drawing by Trautschold, 1842

4–Keller; 5–Dr. Will; 6–Strecker; 7–Aubel (famulus); 9–Warrentrapp; 10–Scherer; 13–A. W. Hofmann

first acquired his interest in chemical phenomena and what he afterwards referred to as his "visual memory." While borrowing chemical books for his father he also obtained access to the court library and devoured all the books bearing on chemistry which he could find there, taking them, as he afterward related, in the order they happened to stand upon the shelves. At the age of sixteen Liebig was apprenticed to an apothecary, and while he soon mastered the chemical side of the business he suffered a misfortune like that which Davy underwent in a similar position. As a boy he had learned from watching the traveling showmen how to prepare silver fulminate, a substance which long had a peculiar fascination for him. Experimenting with it in his new quarters he brought on an explosion which is said to have removed a portion of the roof, and is known to have removed Liebig from the business. He next besought his father to send him to the university. Means were found for this and accordingly in 1820 he matriculated at Bonn, but a year later followed his teacher Kastner (whom after all he did not find very satisfactory) to Erlangen. Here he joined one of the student societies which came under the ban of the government on account of its political tendencies and in consequence he found it prudent to abandon the university. By this time also he was convinced that he could not get such instruction in chemistry as he wanted in Germany, and he applied for a traveling scholarship from the Hessian government in order to continue his studies in Paris. After some difficulties this was granted in 1882. In Paris his abilities, combined with some good fortune, brought Liebig into pleasant relations with Alexander von Humboldt who then spent most of his time in Paris and especially devoted himself to promoting the interests of young men of great promise. By Humboldt he was introduced to Gay-Lussac who admitted him to his laboratory, where he carried out an investigation upon fulminic acid which still commands interest. Meantime the University of Erlangen had conferred the doctor's degree upon Liebig and in 1824 he returned to Germany as professor in the small university of Giessen. The death of the only other professor in the department soon after left him in full charge but the salary was small and all chemical facilities of the worst.

The Laboratory at Giessen.—Finally a deserted barracks was secured and in this was organized what was practically the first laboratory for general instruction in chemistry. A course of study was adopted which has in a measure served as a model for all laboratories of instruction ever since. The student was thoroughly drilled in qualitative and quantitative analysis, prepared some organic compounds and then carried out an investigation suggested by the professor in charge. Despite its limitations the laboratory soon became famous on account of the brilliant researches which proceeded from it, and the inspiration of Liebig's teaching, so that students flocked to it from all over the world. Liebig's own work covered an astonishing range of topics and was at first devoted almost entirely to pure organic chemistry, though later his interest turned with especial favor to agricultural, physiological and food problems. Hence his association with the famous "beef-extract" by which he is still probably best known to the non-chemical public. Liebig's health was much affected by the strenuous efforts of his early career, and laboratory instruction became such a burden to him that in 1852, when called to the professorship at Munich, he accepted only on the condition that he should be entirely relieved of work of this character—a peculiar attitude for the man who had introduced laboratory instruction into Germany. Liebig died at Munich in 1873.

Organic Analysis.—One of his earlier chemical investigations had to do with perfecting the methods of organic analysis. We have already seen that Lavoisier had made important beginnings along this line, and Berzelius and Gay-Lussac had also added improvements, but Liebig gave to the organic combustion practically its present form, and is said to have boasted with characteristic hyperbole that he had so simplified the process that any intelligent monkey should now be able to conduct it successfully. Most students, however, are of the opinion that, if he spoke truly, there is something wrong with the Darwinian theory. The personality of Liebig is an extremely interesting one. He represents what the Germans call a *Feuergeist*, eager, enthusiastic, combative, willing to sacrifice himself (and everyone else) in the pursuit of truth, and inspiring all who surrounded him with the same zeal. He demanded the uttermost of his students and assistants (something for which they thanked him in later years) and he had little patience with anyone who would not stand up for his opinions with an energy akin to his own. It is related that having listened to the praise of someone whom he disliked he finally interjected: "He may be a good man for all I know, but he gives me only a cotton-wool resistance."

Friendship of Liebig and Wöhler.—Such a temperament makes all the more interesting Liebig's remarkable friendship for Wöhler whose nature was the antithesis of all this. Their acquaintance began with a controversy. About 1823 Liebig, who had just analyzed silver fulminate, found that it had the same composition which Wöhler had assigned to the cyanate. That two different substances should have the same composition was then something unheard of, and Liebig, with characteristic self-confidence, declared Wöhler's analyses incorrect. A personal interview not long after led to the repetition of the analyses and Wöhler's vindication, for Liebig seldom allowed his prejudices to blind him to an experimental fact, and here a discovery of the first magnitude was involved, for this was the first case of *isomerism.* The word, however, was not used till 1830 when Berzelius applied it to the relation between tartaric and racemic acids, a case of finer isomerism than he himself realized. Soon after clearing up this point Liebig and Wöhler undertook in collaboration some important investigations in organic chemistry and this gradually brought about the warmest personal attachment. They exchanged frequent letters as long as they lived and it is fortunate that these have been preserved and published. They are interesting on the scientific side because they show us in an entirely informal way how certain problems came to be studied and the manner in which they were attacked. On the human side also they illuminate for us two interesting personalities, both men so thorough, so conscientious, unselfishly devoted to the cause of science and to the truth, but in all else so different—Liebig running over with enthusiasm, irritable, keen for conflict, finding no language quite strong enough to express his feelings, while Wöhler is all gentleness and peace, cautiously avoiding the mildest over-statement, yet gifted with keen insight and full of sly humor which he artfully employs to moderate the turbulence of his friend.

From 1831 Liebig was the most influential editor of the *Annalen der Chemie und Pharmacie*, which acquired its great prestige under his leadership, and he considered it one of the chief duties of the editorial office to defend the truth by pointing out to all poor workers and slovenly thinkers the error of their ways. This course earned him as much gratitude as a similar attitude toward his contemporaries did for Socrates. He made hosts of enemies and became involved in bitter controversies. These too often led him to passionate outbursts which all Wöhler's gentle counsel was unable to restrain, even when expressed as beautifully as in the following passage:

"To make war upon Marchand (or anyone else for that matter) is of no use. You merely consume yourself, get angry, and ruin your liver and your nerves—finally with Morrison's Pills. Imagine yourself in the year 1900, when we shall both have been decomposed again into carbonic acid, water and ammonia, and the lime of our bones belongs perhaps to the very dog who then dishonors our grave. Who then will care whether we lived at peace or in strife? Who then will know anything about your scientific controversies—of your sacrifices of health and peace for science? No one: but your good ideas, the new facts you have discovered, these, purified from all that is unessential, will be known and recognized in the remotest times. But how do I come to counsel the lion to eat sugar!"

It was indeed a hopeless task, as we may see from Liebig's reply to a similar appeal to spare Mitscherlich.

"Poggendorff is a fool, *mon cher*, and even you are half a one with all your representations, which I nevertheless take in good part because I know they are well meant. ——— now knows what he had need to be told—and trembles. That is enough. All the bile which had been long concentrating in me on his account I have now poured out upon him, and I feel relieved to know that the miserable half-way relationship has become clear open enmity. No one is more willing than I to acknowledge a goat when I happen to have shot one, but on the other hand I am bound to defend my convictions to the very death. That and no more I have done."

Dumas.—Jean Baptiste André Dumas was born in Alais in 1800 and began active life as an apothecary in Geneva. Here he had the benefit of associating with the scientific men of the city including Pictet and De la Rive. He early interested himself in physiological problems and even at this time did some work upon the chemistry of the blood which was hitherto unsurpassed, and which called forth highly favorable comment from Berzelius. It also attracted the attention of Alexander von Humboldt who, having occasion to pass through Geneva, took pains to look up Dumas, and, being impressed with his talents, urged him to come to Paris, assuring him that he would find a better scientific atmosphere. Dumas took this advice in 1823 and so began his career in the metropolis at the same time as Liebig and under very similar auspices. Dumas's success in Paris was immediate and complete, and we soon find him installed as a teacher at the *Athenaeum* and later at the *Sorbonne*, as well as giving instruction in other institutions. Dumas was a superior experimenter and clear thinker with an unusual gift of exposition and an imagination which led him to bold and original generalizations. These qualities not infrequently made him a thorn in the side of men like Liebig and Berzelius who doubtless excelled him in thoroughness but were not quite his equals in brilliancy. As a result he almost always emerged without serious harm from the not infrequent controversies in which they tried to overwhelm him by force of accumulated facts.

We now associate Dumas's name with our usual method for the determination of nitrogen in organic compounds, with a method for ascertaining vapor densities and with an experimental determination of the oxygen hydrogen ratio which was a model of accuracy for its time. He also carried on numerous studies in organic chemistry especially such as had to do with the phenomena of substitution and to these we shall have occasion to refer later on. After 1848 Dumas's teaching and experimental work was much interfered with by his devotion to questions of public service such as education, public health and the like. At one time he was a member of the cabinet and he served on numerous commissions. He died at Cannes in 1884.

Organic Chemistry in 1825.—It would be difficult to describe the state of organic chemistry in 1825 from any entirely consistent point of view. Certainly no such point of view then existed. Many important facts were known but all generalizations were extremely vague and unsatisfactory. Lavoisier, indeed, had extended his theory of acids to those of the organic field. He considered these acids to be oxides of *compound* radicals as distinguished from the simple radicals (elements) of the inorganic world. No generalizations were made concerning these radicals though it was assumed that they contained carbon and hydrogen. Other elements, however, were by no means excluded, and when Gay-Lussac studied the cyanogen compounds in 1815 he applied the term to the CN group. Ammonium, too, had been known as a radical since Davy's time. By 1825, also, the general chemical character of alcohols, ethers and esters (or compound ethers as they were then called) was fairly well understood, but all these compounds received formulæ which now seem confusing on account of certain theoretical considerations to which we must next devote our attention.

According to the dualistic system the formula of an 'anhydrous acid' was best fixed by deducting from the formula of a salt that of the base; thus in calcium sulphate CaO,SO_3 the acid is SO_3. It followed that in calcium acetate $C_4H_6O_4Ca$ the acid was to be considered as that which combined with the lime, namely $C_4H_6O_3$. The fact that glacial acetic acid contains a molecule more water than this was disregarded, and it became the habit when an acid was analyzed to throw away enough oxygen and hydrogen from the formula to account for whatever water was 'lost' in salt formation. If this water could be easily expelled from the free acid by heat it was called water of crystallization, if not, the name applied was 'water of composition' or some other term of as little significance. The vapor density was allowed to have no influence in the determination of the formula because, since the rejection of Avogadro's ideas by Dalton and Berzelius, most chemists had adopted an attitude of reserve as to the significance of this property. If we accept the above formula for acetic acid, then ethyl alcohol (from which it is formed on oxidation) naturally becomes $C_4H_{10}O_2$ and ether which can be formed by dehydrating alcohol appears as C_4H_8O. Similarly marsh-gas was usually written C_2H_8 and ethylene C_2H_4. The foregoing still fails to give a complete idea of the existing confusion, for many chemists whose theories followed those of Wollaston (page 80) were using the equivalents $C = 6$ and $O = 8$. Still others more eclectic in taste preferred $C = 6$ and $O = 16$ and wrote their organic formulæ on this basis.

We may concede at the outset that in 1825 it would have been utterly impracticable to determine the constitution of an organic compound in the sense in which we now employ that expression. The false view of the nature of acids, however, which has been described above, and which was imposed upon organic chemistry from without, solely in the interest of a consistent system, blinded chemists to obvious and simple relationships and prevented them from doing even what they might toward a natural systematization of the facts. We shall see that this unfortunate tendency was destined to do still further harm in the future, and there is no more instructive example than this of what a pernicious theory can sometimes do toward obstructing the healthy progress of science.

Original Attitude of Berzelius.—Although the ideas outlined above were essentially dualistic, Berzelius did not at first make any serious attempt to emphasize this, or to apply his system at all generally in the organic field. A wise caution led him to point out that the organic compounds were all products of the animal or plant organism. He therefore ascribed their existence as well as their original formation to the *vital force* and freed them for the present from the tyranny of his electrochemical rules. In 1828, however, Wöhler made a discovery which cut off hope that the issue could much longer be avoided in this way. He treated potassium cyanate with ammonium sulphate in the hope of obtaining ammonium cyanate but the solution on evaporation yielded instead urea! The cyanates were at that time classed as inorganic compounds and not long after they were prepared from the elements, so that the complete synthesis of one well-known organic product of the animal organism was an accomplished fact. Other syntheses followed, and it soon became evident that the assumption of a 'vital force' was an untenable hypothesis. Organic compounds like inorganic must owe their existence to chemical affinity, but concerning the nature of chemical affinity Berzelius was already committed. It was a manifestation of "the two opposing forces of positive and negative electricity * * * since there is no third force" (see page 113). To apply such a theory to organic compounds it was necessary to think of these as composed like salts of a positive and a negative component, and since in the majority of cases organic compounds are not electrolytes, the nature and composition of the components could only be ascertained by bold assumptions based on other facts of chemical experience.

The Etherin Theory.—Berzelius was as yet by no means ready to indulge in such speculations when in 1828, the same year as Wöhler's discovery, Dumas made the suggestion that a considerably better insight into the chemistry of many substances associated with ordinary alcohol might be attained if they were all considered as addition products of ethylene. We shall employ modern atomic weights in illustrating his views for Dumas was now using the atomic weights C = 6 and O = 16, hence his formulæ appear needlessly confusing to the modern eye.

Dumas pointed out that the substances we now term ethyl halides might be advantageously formulated as addition products of ethylene and halogen acids; alcohol, of ethylene and water; while ether was a compound of ethylene with less water; ethyl acetate, of ethylene, water and acetic acid; and ethyl sulphuric acid was an addition product of ethylene, sulphuric acid and water, as indicated in the following table:

	Modern formula	Dumas's formula
Ethyl chloride	C_2H_5Cl	C_2H_4+HCl
Alcohol	C_2H_5OH	$C_2H_4+H_2O$
Ether	$(C_2H_5)_2O$	$2C_2H_4+H_2O$
Ethyl acetate	$CH_3COOC_2H_5$	$C_2H_4+C_4H_8O_4+H_2O$
Ethyl sulphuric acid	$C_2H_5SO_4H$	$C_2H_4+SO_3+H_2O$

To us, these formulæ appear somewhat unnatural but it was possible to support them by a good deal of experimental evidence. They explained fairly well the formation of ethylene and ether by the action of dehydrating agents upon alcohol; the formation of ethyl acetate by the action of the acid upon alcohol, and of ethyl sulphuric acid by the action of sulphuric acid upon alcohol or upon ethylene.

The theory had another advantage which appealed strongly to Dumas. It represented ethylene as analogous to ammonia. We have already seen how the marked difference between the latter substance and other bases had induced Davy to seek for oxygen in its composition (page 93), and led Berzelius to doubt even the elementary nature of nitrogen (page 111). According to Dumas, however, ammonia now found its natural place among the organic radicals. Ammonium chloride, NH_3+HCl, was comparable with ethyl chloride, ammonium acetate, $2NH_3+C_4H_6O_3+H_2O$, with ethyl acetate and so on. Dumas became so enamoured with this feature of his theory that he declared ethylene a true base which *would* turn litmus blue *if* it were only soluble in water!

Benzoyl.—Berzelius received the new ideas with cautious reserve. They were essentially dualistic in spirit but they failed to emphasize that importance of oxygen which was the vital point in his system. He commented therefore to the effect that Dumas had found an interesting and suggestive way of symbolizing the relationship of the compounds mentioned, but he expressed no faith that the latter were really so constituted. The theory was, however, soon to receive support from another quarter. In 1832 Liebig and Wöhler published their justly famous paper upon the oil of bitter almonds. As we know this material consists essentially of benzaldehyde, the first substance of this important class to receive thorough study. Liebig and Wöhler observed its oxidation to benzoic acid, its transformation to benzoin, and by the action of chlorine they obtained benzoyl chloride and from this by double decomposition the bromide, iodide and cyanide, as well as benzamide and ethyl benzoate. The theoretical results of the investigation may be summed up in the statement that in all these compounds they found evidence for the presence of a common radical, $C_{14}H_{10}O_2$, which they named *benzoyl.* It was, as we see, our modern benzoyl whose formula, however, had been doubled from theoretical considerations like those already discussed. To Liebig and Wöhler benzaldehyde was an addition product of this radical with hydrogen, $C_{14}H_{10}O_2+H_2$; benzoic acid with oxygen, $C_{14}H_{10}O_2+O$; benzoyl chloride with chlorine, $C_{14}H_{10}O_2+Cl_2$, and so on. The close relationship of all the compounds with each other was indisputable, and Berzelius was so carried away by the brilliant achievement that he wrote to Liebig and Wöhler a most enthusiastic letter in which he suggested the use of the name *Orthroin* for the new radical, to show that its discovery meant the dawn of a new day for the science, and at the same time he confessed his own belief that ethylene was the true radical in the alcohol group, and suggested that it henceforth be called *Etherin.* For this reason Dumas's original suggestion is known as the Etherin Theory to this day.

Berzelius and the Ethyl Theory.—The enthusiasm of Berzelius was short-lived. The etherin theory did not bring out the importance of oxygen in the way which his system demanded. Lavoisier had defined a compound radical as a group of elements which behaves like a single one, and unites with oxygen to form an acid. To Lavoisier these expressions meant essentially the same thing, for to him the chief function of any element was to unite with oxygen. Of late years, however, that part of the definition was being lost sight of. In cyanogen, in ammonium, and now in etherin what was being emphasized was the permanence of the group, not its union with oxygen. As early as 1833 Berzelius resolved to maintain this latter point at all costs. A radical was to him that which unites with oxygen. It therefore could not contain oxygen. This decided him to break with the theory of etherin and benzoyl, and set up other radicals which would fit better into his system. He sought an occasion for this in a question of little intrinsic importance. The etherin theory formulated the barium salt of isethionic acid as $2C_2H_4+2SO_3+BaO+H_2O$, and that of ethyl sulphuric acid as $2C_2H_4+2SO_3+BaO+2H_2O$. Inasmuch, however, as one salt did not go over into the other by boiling with water it could not, he explained, contain water ready formed. He therefore assumed a new radical *ethyl*, C_4H_{10}. The oxide of ethyl was ether, $C_4H_{10}O$, and ether might be considered as uniting with anhydrous acids to form esters just as metallic oxides united with them to form salts. Ethyl acetate was a binary addition product of ether and acetic acid, $C_4H_{10}O+C_4H_6O_3$, entirely analogous to calcium acetate, $CaO+C_4H_6O_3$. Benzoyl, also, he discarded as a radical. He now regarded it as the oxide of a true radical, $C_{14}H_{10}$, of which benzoic acid is a still higher oxide. Ammonia also found a place in the system, for in accordance with the line of argument developed on page 111 its analogy with other bases can be preserved by assuming that in salt formation ammonia takes up the superfluous water of the acid to form ammonium oxide, which then adds directly to the acid. On this basis ammonium sulphate is to be written $N_2H_8O+SO_3$, ammonium nitrate, $N_2H_8O+N_2O_5$, and ammonium acetate, $N_2H_8O+C_4H_6O_3$.

The fundamental idea will perhaps appear more plainly if some of the characteristic formulæ are tabulated:

Ethyl, C_4H_{10}	Calcium acetate, $CaO,C_4H_6O_3$
Ether, $C_4H_{10}O$	Ethyl acetate, $C_4H_{10}O,C_4H_6O_3$
Alcohol, $C_4H_{10}O,H_2O$	Ammonium oxide,[1] $(NH_4)_2O$
Acetic acid,[1] $C_4H_6O_3$	Ammonium sulphate, N_2H_8O,SO_3
Glacial acetic acid,$C_4H_6O_3$, H_2O	Ammonium acetate, $N_2H_8O,C_4H_6O_3$

Liebig adopted these views, pointing out that alcohol could be considered as a compound of ether and water, $C_4H_{10}O+H_2O$. There resulted a long controversy between Liebig and Dumas into the details of which we have no occasion to enter. Dumas, however, acknowledged his conversion in 1837, and the two chemists agreed to collaborate henceforward in their studies in organic chemistry. This 'era of good feeling' proved to be only the moment of calm before the storm, but the situation has interest because it represents the last great triumph of Berzelius. For the moment a dualistic system essentially electrical, based upon the combination of positive and negative elements or radicals with oxygen, held practically undisputed sway in both organic and inorganic chemistry. Such a condition, however, could not last.

Liebig's Acetyl Theory.—By 1839 Liebig had again begun to modify his views. Some time before, Regnault had treated ethylene chloride with alkali and obtained chloroethylene which he formulated, $C_4H_6Cl_2$. This contains too little hydrogen to be an etherin compound, and this naturally suggested the adoption of C_4H_6 as a radical. Liebig accepted it as such and named it *acetyl* because acetic acid, $C_4H_6O_3$, could be considered as its oxide. By its use, also, acetic acid and ordinary alcohol could be formulated from a common point of view, for etherin itself could now be regarded as a compound of acetyl with hydrogen, while ethyl was either a compound of etherin with hydrogen or of acetyl with more hydrogen. Liebig pointed out with great satisfaction that this latest of his theories was one upon which the adherents of both etherin and ethyl could now compromise in harmony. He seems hardly to have realized that to attain this formal harmony he had sacrificed almost all the principles involved in the idea of radicals.

[1] Hypothetical.

Review.—The theoretical conceptions which have just been outlined are commonly grouped together under the name of the first Radical Theory. This had begun with etherin and one of its most fundamental ideas was the reality of the radicals. They were supposed to preëxist in the compound, and in a measure determine its chemical properties by their own independent behavior. This made it at first a fundamental tenet of the creed that the radicals should be capable of existence in the free state. This was true of ethylene and the existence of free cyanogen and free cacodyl was considered important evidence that these complexes were true radicals. When etherin was exchanged for ethyl it was for the sake of consistency, to make the theories of organic and inorganic combination similar in form but, as Dumas pointed out, it involved the sacrifice of a real compound for a hypothetical group. Finally, when ethyl gave place to acetyl it was merely in the interest of harmony and convenience. The radical had now become something artificial, and the word was beginning to acquire its modern meaning, namely, a number of elements grouped together for convenience in tracing genetic relationships. As much as this would probably not have been admitted by its adherents at this time, but the theory had nevertheless purchased flexibility at the cost of significance, and no one except Berzelius any longer preserved enough faith in the reality of its fundamental principles to defend it efficiently from the vigorous attack which was about to be launched against it by Dumas.

Literature

The personal elements in the discussion of this period are best brought out in the letters of the participants. The most interesting collections are the following: *Mitscherlich, Gesammelte Schriften*, Berlin, 1896, which contains some interesting letters by BERZELIUS; *Briefwechsel zwischen Berzelius und Wöhler*, edited by WALLACH, Leipzig, 1901; and *Briefwechsel zwischen Liebig und Wöhler*, edited by HOFMANN, Leipzig, 1888. An interesting collection of youthful letters by Wöhler has also been published by KAHLBAUM under the title *Friedrich Wöhler, ein Jugendsbildniss in Briefen*, Leipzig, 1900.

Liebig's character attracted many biographers. The authoritative life is that by VOLHARD, Leipzig, 1904. A much shorter one was published in English by SHENSTONE, New York, 1895. Liebig's character and temperament are also thoughtfully discussed in OSTWALD's *Grosse Männer*. HOF-

MANN's lecture before the Chemical Society of London in 1875 was separately published in the following year under the title *The Life and Work of Liebig.* HOFMANN included a German version of the same address in his *Zur Erinnerung an Vorangegangene Freunde* (3 vols.), Braunschweig, 1888. This collection also contains commemorative addresses on Wöhler and Dumas. The two latter are discussed in THORPE's *Essays.*

As pointed out in the last chapter the scientific work of the period is best studied in the journals. For those to whom these are not accessible a comprehensive discussion can be found in the pages of KOPP and LADENBURG already alluded to. See also OSTWALD's *Klassiker* No. 22 for the paper of Liebig and Wöhler on the radical of benzoic acid.

CHAPTER XIII

THE REACTION AGAINST BERZELIUS

Substitution in Organic Chemistry.—Hofmann vouches for the tradition that Dumas's interest in substitution began when a great social function at the Tuilleries was spoiled by the choking fumes emitted from the candles. These were turned over to Dumas for investigation, who found that they had all the superficial appearance of ordinary candles but emitted clouds of hydrochloric acid when lighted. It proved that the wax from which they had been made had been bleached by chlorine and this evidently had entered into its chemical composition. Dumas soon after proceeded to treat many other organic compounds with chlorine and bromine, and he found a common if not universal result, that in such cases more or less hydrogen was substituted by an equivalent quantity of halogen. It also frequently happened that this exchange caused surprisingly little change in properties.

The Nucleus Theory.—About 1836 Dumas's countryman Laurent took up the idea and developed it into a flexible and comprehensive system which came to be called the Nucleus Theory. It was essentially a radical theory in which, however, new radicals could be formed by substitution whenever this was convenient, so that it could readily be made to cover almost any possible cases. As a system of classification this had many merits, and it was adopted for this purpose in Gmelin's great *Handbuch.* We have, however, no occasion to discuss its details, for it was frankly artificial and never received any recognition by the great chemical authorities. Liebig attacked it vigorously on the experimental side, Berzelius denounced it even more bitterly as a theory, and Dumas largely ignored it because he was himself about to take up a somewhat similar position.

First Type Theory.—In 1839 he treated acetic acid with chlorine and obtained trichloroacetic acid. In spite of the great

JEAN BAPTISTE ANDRÉ DUMAS
1800–1884

[Facing page 13/

rence in composition the new acid resembled acetic in a ing degree. It had the same basicity, and when distilled alkali one yielded chloroform while the other yielded marsh-showing that these two substances also stood to each other he same relation as the two acids. Upon these reactions has based what later became known as the first Type Theory. his he distinguished on the one hand *chemical types* to which nged substances closely resembling each other like chloroform bromoform, and on the other, *mechanical* types where the larity was more formal but the relationship was still one of stitution. The following list illustrates the latter class.

Marsh-gas	$C_2H_2H_6$
Formic acid	$C_2H_2O_3$
Chloroform	$C_2H_2Cl_6$
Carbon chloride	$C_2Cl_2Cl_6$

ill be seen that by force of habit the formulæ are still written dualistic manner but the dualistic spirit is entirely absent. nas compares the relation of atoms in a compound to that of planets in the solar system. Compounds according to his are built upon a chemical type, and their properties depend n the number of the atoms making up the type together with r relative position. The nature of the atoms themselves is ar less consequence. Dumas points out with much feeling in all the radical theories it had hitherto been an unprofitable ssity to divide every compound into two parts whether hing in its chemical behavior called for such a division or not. him henceforward every compound is one unit, and while rical forces may be involved in its formation, there is no t dualism, and no fixed charges of electricity belonging to icular atoms are involved.

ie instances of substitution continued to multiply and the theory became popular. Dumas, however, in his enthusiasm ied it with a freedom which alarmed the more thoughtful. was anxious to see substitution in every reaction, he recog-d not only substitution of hydrogen in the types, but also of gen, oxygen and even of carbon—all without changing the . The substituent also might be not only an element but a ip, so that in the eyes of Dumas compounds became associated re no one else could see a relationship. These excesses came near bringing the whole idea into ridicule, and in 1840 an article was published in the *Annalen* ostensibly from the pen of a certain S. C. H. Windler[1] in which the author describes the remarkable results which he has obtained by treating manganese acetate with chlorine. In this way he substituted first the hydrogen, then the oxygen, then the carbon, and finally the manganese, and so obtained a product similar to the original acetate but which consisted entirely of chlorine and water. He then goes on to recommend for use as night-caps certain fabrics, which he says may be had in Paris, which have all the properties of cotton, though they consist entirely of chlorine!

Despite such good-natured attempts at satire most of the leading chemists of the time were seriously convinced that there was much which was sound and profitable in the new views, although, as might have been expected, the attitude of Berzelius was irreconcilably hostile and bitter.

Attitude of Berzelius.—When Laurent originated the nucleus theory, Berzelius had condemned it without mercy because it involved the substitution of hydrogen by chlorine in the radicals. For him the halogens and sulphur were *negative* elements, and, while they might sometimes replace oxygen in the electro-negative portion of a complex, just as one metal replaces another in a series of salts, yet that halogen should replace hydrogen in a positive radical without changing the chemical nature of the latter was unthinkable. It goes without saying that Dumas's views which rejected dualism altogether, were to Berzelius nothing less than unspeakable heresy. So far as trichloroacetic acid was concerned, he denied everything which could be denied, maintaining as long as possible that this acid had no similarity to acetic acid. When, however, in 1842 Melsens succeeded in reversing the substitution and passing back from the chlorinated compound to acetic acid the analogy could no longer be disputed. Berzelius then took refuge in a new formula for acetic acid. It was no longer a simple oxide of C_4H_6 but a "conjugate" compound of oxalic acid C_2O_3 combined with a group C_2H_6. This latter was called the *copula*, and in this copula substitution in the sense of Dumas and Laurent might take place—apparently

[1] It was written by Wöhler.

because it really was not the seat of the acid properties of the compound:

Acetic acid	C_2O_3,C_2H_6
Trichloroacetic acid	C_2O_3,C_2Cl_6

This curious attempt to keep half a molecule dualistic by sacrificing the other half, contained an idea of which Kolbe was able to make valuable use later. At the time, however, it was generally regarded as only a makeshift designed to save acknowledgment of defeat.

In 1837 Liebig and Dumas dealt a blow to another favored theory of Berzelius. This time the attack was upon the oxygen theory of acids, and the occasion was furnished by certain researches of Graham published about four years earlier.

Graham.—Thomas Graham was born in Glasgow in 1805 and graduated from the university there in 1824. After two years spent in the laboratory of J. C. Hope in Edinburgh, he returned to his native city and began teaching mathematics and chemistry, at first privately and then in the Mechanics Institute and the Andersonian Institution. In 1837 he was called to University College in London, and in 1855 succeeded Sir John Herschel as master of the mint, a position which he retained till his death in 1869. As early as 1829 he had already begun the study of diffusion in gases which led to the discovery of his famous law, that the rate of diffusion is inversely proportional to the square root of the density. He next proceeded to study the diffusion of liquids and here his researches laid the foundations of our knowledge of osmosis and drew the distinction which we still make between crystalloid and colloid solutions. Indeed he is justly regarded as the founder of colloidal chemistry. The work upon phosphoric acid with which we are immediately concerned was published in 1833.

The Polybasic Acids.—To the chemist of the twentieth century there are few things harder to realize than that in 1833 all acids were considered as monobasic, even sulphuric, oxalic and carbonic. That these acids are dibasic seems well-nigh self-evident to us, chiefly because we think of the existence of acid salts, but we have to remember in the first place that monobasic acids, notably hydrofluoric, frequently form acid salts, and furthermore that the dualistic system formulated these compounds in a manner entirely out of harmony with the modern point of view. The neutral and acid sulphates of potassium, for example, were written K_2O,SO_3 and $K_2O,2SO_3$ respectively, and the latter substance was called the *bisulphate* because it represents the union of the base with twice as much acid as in the neutral salt. It is true that this disregarded the additional water for which our modern formula $KHSO_4$ accounts, but in accordance with a point of view with which we are already familiar this was regarded as something akin to water of crystallization. In the same way there were bicarbonates $K_2O,2CO_2$, bichromates $K_2O,2CrO_3$, and so forth, whose names still persist colloquially, though this last example, where there is no water, well shows how the dualistic theory also concealed the difference between acid and pyro-salts.

Another difficulty lay in the fact that many chemists at this time doubled the atomic weights of the alkali metals, writing KO,SO_3, NaO,SO_3, as well as CaO,SO_3. This confused utterly the distinction between the monovalent and bivalent metals, though any allusion to valence is really misleading in this connection, for the conception simply did not exist, and there were then no data on which it could have been built up. Chemists were by no means agreed as to whether water should be written HO or H_2O and in such a state of things to speak even of the valence of oxygen is an absurdity.

When Graham took up the study of phosphoric acid two phosphates of soda were recognized. One was the salt we now call the pyrophosphate, $Na_4P_2O_7$, and the other the ordinary mono-acid phosphate Na_2HPO_4. It is well to remember that the latter is neutral to most indicators. Their distinct individuality was shown by the fact that in solution one gave a white precipitate with silver nitrate, and the other a yellow, and Berthollet had observed that, in the latter case, the solution became acid after precipitation, an apparent exception to Richter's rule (page 64). In spite of this difference in properties, both salts, if we disregard the water, seemed equally entitled to the formula $2Na_2O,P_2O_5$ and the relationship was considered a case of isomerism. Graham's fundamental experimental discovery was that when the ordinary phosphate is heated it loses water, and the residue goes over to the pyrophosphate which now of course gives the characteristic white precipitate with silver nitrate. The difference between the two salts was therefore the molecule of water lost in heating, and this could hardly be water of crystallization, else solutions of the two salts would be identical. Graham went on to prepare the metaphosphate and showed that its relation to the acid phosphate NaH_2PO_4 is analogous to that between the pyrophosphate and the (so-called) neutral one. Furthermore it was found possible to prepare double salts of phosphoric acid which differed in their properties from the phosphates of either metal considered separately, and this also led to the conclusion that here two or more bases are combined in the same molecule, and Graham was able to show that all which he had discovered concerning phosphoric acid held equally true of arsenic acid. Graham concluded that in such acids the essential thing is the water content. To him, therefore, an acid was no longer what it was in the eyes of Lavoisier and Berzelius, the oxide of a non-metal, but rather the compound of such an oxide with a certain quantity of water which he called "basic water." Salt formation consequently consisted in substitution of this basic water by a metallic oxide. These ideas may be made clearer by the following table which shows the composition of a number of common phosphates both in accordance with Graham's view and as we should now formulate them.

	According to Graham	Modern formula
Phosphoric acid	$P_2O_5,3H_2O$	H_3PO_4
Tertiary phosphate of soda	$P_2O_5,3Na_2O$	Na_3PO_4
Ordinary phosphate of soda	$P_2O_5,2Na_2O,H_2O$	Na_2HPO_4
Pyrophosphate of soda	$P_2O_5,2Na_2O$	$Na_4P_2O_7$
Pyrophosphoric acid	$P_2O_5,2H_2O$	$H_4P_2O_7$
Acid phosphate of soda	$P_2O_5,Na_2O,2H_2O$	NaH_2PO_4
Metaphosphate of soda	P_2O_5,Na_2O	$NaPO_3$
Metaphosphoric acid	P_2O_5,H_2O	HPO_3
Microcosmic salt	$P_2O_5,Na_2O,(NH_4)_2O,H_2O$.	$Na(NH_4)HPO_4$

It will be recognized from the above that Graham's conclusions involve no break in the traditional theory of acids. The only novelty is that in the free acids and acid salts "basic water" now plays the rôle of positive component.

Liebig on the Polybasic Organic Acids.—In 1837 Liebig and Dumas undertook an extension of the work of Graham by studying the polybasic organic acids. It will be recalled that they had just concluded an armistice with the declaration that they would henceforward study organic chemistry in collaboration. One brief paper on this subject was the only one so published before relations again became strained, but in the following year Liebig went on with the work alone, and after studying the salts of such acids as citric, tartaric, cyanuric and muconic, came to the conclusion that these were true polybasic acids. The criterion by which Liebig decided whether or not he had to do with a polybasic acid was the formation of salts containing two or more bases; thus if tartaric acid is neutralized with a mixture of soda and ammonia a double salt is formed unlike either sodium or ammonium tartrate. Liebig considered this evidence that tartaric acid must neutralize two atoms of base. His conclusion was of course correct but his method was in a measure faulty for it misled him in the case of sulphuric acid. When this acid is neutralized with an equivalent mixture of potash and soda a mixture of the two sulphates is obtained. Liebig therefore continued to consider sulphuric acid monobasic. There were other minor errors of the same kind in the work, but these sink into insignificance in comparison with the far-reaching general conclusions which Liebig was able to draw from it. In the first place he abandoned the theory of composition which Graham had so ingeniously applied to phosphoric and arsenic acids. In these special cases it was possible to obtain the salt of one acid from that of another by merely driving off water. In the organic field we can see how difficult it would be to apply the same reasoning, if we think of the complex reactions which take place when tartaric or citric acids are distilled. When the water of crystallization has been removed from an organic acid there is no way in which any special form of water it may contain can be distinguished from any other atoms of oxygen and hydrogen in its composition, nor is there any reason to assume that it contains any, except for the sake of theoretical uniformity. In fact, Liebig was fast becoming weary of all these different hypothetical forms of water with which the dualistic theory had loaded organic compounds, and he decided that it was far simpler and more satisfactory to discard the oxygen theory of acids, and define these substances as compounds containing *hydrogen* which can be replaced by a metal. It will be recalled that Davy had made a move in the same direction long before, but the opposition of Berzelius had limited the application of the idea to those acids known to contain no oxygen, and Berzelius had denied the elementary nature of chlorine to the last possible moment in order to maintain uniformity in the theory of acids. He argued with much force that if the existence of hydrogen acids is admitted, then the action of sulphuric acid upon magnesia for example:

$$SO_3,H_2O + MgO = SO_3,MgO + H_2O$$

becomes entirely different from that of hydrochloric acid upon the same base:

$$2HCl + MgO = MgCl_2 + H_2O$$

but he finally accepted the contradiction rather than secure uniformity in the only logical way, by surrendering the oxygen theory of Lavoisier. Liebig, however, now took this step and expressed himself in words of characteristic energy:

"In order to explain one and the same phenomenon we use two sets of forms; we are forced to assign to water the most manifold properties; we have basic water, halhydrate water,[1] water of crystallization; we see it enter compounds where it ceases to exercise any of these functions, and all this for no other reason than that we have drawn a distinction between haloid salts and oxygen salts, a distinction which we do not observe in the compounds themselves. They have similar properties in all their relationships."

As might have been expected Berzelius protested with great vigor, but Liebig's opinions carried the day, and though chemists long continued in their writings to formulate the salts of oxygen acids in the old way, they did so henceforward more from force of habit than conviction. The old point of view had become intrenched in the nomenclature, and occasionally crops up even in the literature of the present day.

It should be emphasized that Liebig's discoveries were not en-

[1] Liebig applies this term to the "basic water" of acid salts such as $P_2O_5,2Na_2O,H_2O$.

tirely incompatible with a dualistic conception of acids and salts. The former might still be written as compounds of hydrogen with a radical like SO_4, but to attempt this involved as complete a break with the historical associations of the idea as to surrender it altogether, and the tendencies of the time all led in the latter direction. Unitary conceptions were fast replacing the old dualistic ones. We have seen how Dumas was already evolving the idea that organic compounds are built upon unitary types and are related to each other by substitution. This word became the shibboleth of the day and salts began to be defined as acids in which hydrogen is *substituted* by a metal.

Vapor Density as a Measure of Molecular Weight.—These attacks were by no means the first to which the system of Berzelius had been subjected. It will be recalled that the fundamental criterion which had guided him in the selection of atomic weights had been the law of combining gas volumes originally suggested by Gay-Lussac, and which Berzelius had adopted in the form that equal volumes of the *elementary* gases contain the same number of atoms. About 1826 Dumas became interested in a line of reasoning similar to that of Avogadro, and in order to test this experimentally he devised the method of determining vapor densities which we still associate with his name. The results were a disappointment. He had expected to find strict proportionality between the vapor density and combining weight, but when he came to vaporize mercury and sulphur he obtained values incompatible with so simple a hypothesis. He might of course have assumed as we do today that the number of atoms in the molecule of an elementary gas is a constant which is characteristic for the element concerned, but at this time such an assumption seemed justified by no other independent evidence and Dumas drew instead the alternative conclusion that vapor densities were an unreliable guide in the determination of molecular weights. Berzelius, too, felt constrained to agree with him in so far as to limit the application of Gay-Lussac's hypothesis to those elements which are gaseous under ordinary conditions. In this form, of course, the generalization had become so limited in its application as to be well-nigh worthless.

Polymorphism.—It was hardly better with the other criteria, the laws of isomorphism and of atomic heats. As specific heats

Thomas Graham
1805–1869

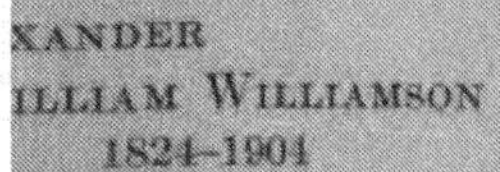

xander
illiam Williamson
1824–1904

oduced from Thorpe's
tory of Chemistry" by
ind permission of G. P.
am and Sons.

Michael Faraday
1791–1867

ere determined with greater accuracy it became increasingly vident that the law of Dulong and Petit was inapplicable to ie elements of low atomic weight, at least within the ordinary inge of temperature. Furthermore the discovery of poly-orphism by Mitscherlich did much to weaken the theoretical inclusions which had been drawn from his earlier discovery of omorphism. If one and the same substance can crystallize in vo or more totally different forms, of what value are speculations ised upon the fact (perhaps only a coincidence) that different impounds sometimes crystallize in forms essentially the same?

Faraday's Law.—Still another difficulty for Berzelius was found Faraday's law. In 1834 Faraday, while seeking additional ridence for the identity of static and galvanic electricity, hit ɔon a standard of comparison in the decomposition of potas-um iodide. When a battery of Leyden jars is discharged irough a piece of filter-paper wet with a solution of potassium dide and starch, a blue stain is produced at the point where the ɔsitive pole touches the paper. It occurred to Faraday to easure the time required by a weak galvanic battery of known rength to produce a stain of the same intensity. He then asoned that the quantities of electricity which had passed must e the same in the two cases. He followed up these observations ith a long series of remarkable experiments in which a great ariety of substances were decomposed by the current, and the uantities of material formed at the electrodes compared with ie amount of hydrogen liberated from a solution of dilute ilphuric acid which was connected in series. In all cases he iund that these quantities were proportional to the combining eights of the substances concerned.

These results convinced Faraday that chemical affinity is lentical with electricity and gave him, as he believed, the best riterion hitherto attained for the determination of atomic veights, for at this time he thought that only such salts are de-omposed by the current as consist of one positive and one iegative atom. The figures obtained on this supposition could iot, however, be harmonized with the atomic weights of Berzelius. In the case of the elementary gases the latter had followed Gay-Lussac in the belief that equal volumes contained an umber of atoms. With hydrogen as unity this wo oxygen the atomic weight of 16 and water the formula H_2O. The current, however, liberates from water eight grams of oxygen for one of hydrogen and in consequence, for Faraday, the formula of water was HO and the atomic weight of oxygen 8. Similar reasoning led to atomic weights for carbon, calcium and some other elements half as great as those of Berzelius, while those of the halogens remained the same. It will be recalled that these figures are essentially the same as the so-called 'equivalents' advocated by Wollaston in 1813. They had found little acceptance on the continent on account of the dominating influence of Berzelius but they had always been popular in England and, from this time on, were destined to come into still more general use, rather to the detriment of the science.

Faraday's Researches in Electricity.—The portion of Faraday's famous *Experimental Researches in Electricity* which we have just briefly sketched was a contribution of the utmost importance to the special field of electrochemistry and smoothed the way for the great advances which were later to be made there. It was at this time, for example, that Faraday introduced the words *anode,*

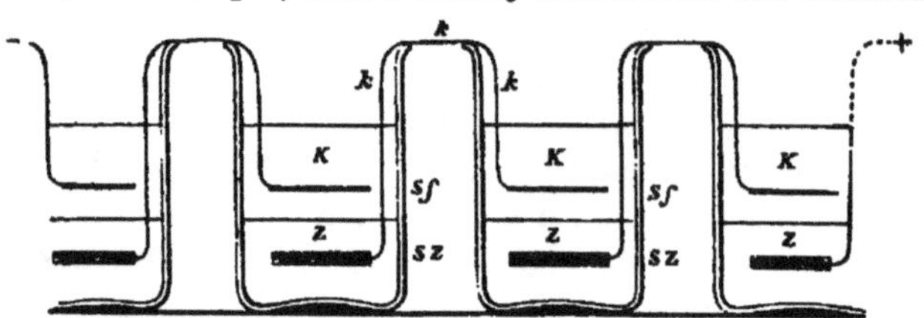

EXPERIMENT ILLUSTRATING THE ARGUMENT OF BERZELIUS FOR THE 'CONTACT' THEORY OF THE ORIGIN OF THE GALVANIC CURRENT

cathode, ion, etc., terms now so important in the theory of electrolytic dissociation. Faraday of course did not grasp the fact that electrolytes become dissociated by the simple fact of solution, nor were his ions quite identical in individual cases with those which we now assume. Nevertheless he did recognize the ion as that portion of a dissolved substance which carries the current in electrolysis. All this served to make the conception helpfully familiar later on.

That which was immediately important was the clearness with which Faraday showed that a certain definite *quantity* of electricity is involved whenever a chemical equivalent is transformed by the current. This threw a new light upon the old controversy concerning the origin of the current itself (page 85) and made it clearer than ever to Faraday that this could only be accounted for as the result of the chemical action in the battery. Such reasoning, however, failed utterly to convince Berzelius, who had some time before gone over to the contact theory, being converted thereto by an experiment which he ever afterward quoted as absolutely conclusive. The reasoning involved is something like this: In a simple cell containing copper, zinc and sulphuric acid, those who believe in the chemical origin of the current must attribute it to the action of the acid upon the zinc. If, now, a cell could be constructed in which the copper was acted upon and the zinc not, then the current should flow in the opposite direction. To test this Berzelius immersed zinc in a solution of zinc sulphate and then, without mixing, carefully introduced above the zinc sulphate a solution of nitric acid in which was finally immersed a plate of copper. At first, of course the copper was attacked, but as soon as the metallic plates had been connected by a wire the current passed *in the customary direction* and the zinc began to go into solution. To the mind of Berzelius this was conclusive evidence that the current is the *cause* and not the *effect* of the chemical action in the battery, and can sometimes even reverse the natural course of such action.

Faraday's law also touched Berzelius in a point where he was still more sensitive, for by implication it seemed to threaten the whole theory of chemical action which he had built up upon the assumption that different electrical charges are carried by the various elements. Faraday, it is true, had clearly pointed out that while the *quantity* of electricity was proportional to the quantity of change, it was the *intensity* of the current required to effect a given decomposition which was the measure of the affinities concerned. Nevertheless, few scientists had hitherto clearly differentiated the two conceptions and Berzelius was certainly now too old to learn. In his text-book he complains that according to Faraday's view

"the same electric current which separates an atom of silver from an atom of oxygen also separates an atom of potassium from an atom of oxygen, whereas the first is one of the loosest and the last one of the firmest combinations which we know."

Berzelius, of course, could have no sympathy with such a view and he attacked it bitterly though to little purpose. The experiments of Faraday were too convincing.

The Combination of Influences against Berzelius.—Bitterness, unfortunately, was destined to be the portion of Berzelius's later years. He had spent his life in a single-hearted devotion to the science which has never been excelled, and had devoted his great talents and tireless energy to organizing and establishing it upon foundations which he believed to be impregnable; yet now at the close of his life he had to see practically every generalization upon which he had set his heart undermined and discredited, while the world was filled with new doctrines which he believed could lead only back to chaos.

In a sense he was right. The newer discoveries had clouded the simplicity and order promised by the older generalizations. The law of isomorphism, for example, had been weakened by the discovery of polymorphism. The law of Dulong and Petit had shown a painful number of exceptions. Vapor densities had proved an unreliable measure of molecular magnitude chiefly on account of dissociation—a phenomenon not understood. The discoveries of Faraday threw into still further doubt all the criteria for the determination of atomic weights, and struck directly at the cherished electrical theory of chemical affinity. The work of Liebig on the polybasic acids displaced oxygen from the sacred place at the center of the chemical system where Lavoisier and Berzelius had done so much to maintain it, and finally—worst and most crushing of all—the theory of electric dualism had broken down utterly in the organic field, and the arch-heretic Dumas was even now setting up in its place a unitary system destined to convince chemists that electricity played no fundamental part in chemical reactions. With a different temperament Berzelius might have possessed his soul in patience resting on the record of his magnificent experimental work, but he valued his theoretical system above everything, and in its defence spent too much of his last years in violent personal attacks upon its enemies. As a matter of fact the collapse of the great system did lead to a period of general scepticism toward the very possibility of far-reaching generalizations, but out of this confusion there gradually rose another system better than the old, in which each important principle championed by Berzelius was destined to find its appropriate place.

Literature

There is a life of Graham by ROBERT ANGUS SMITH, Glasgow, 1884. THORPE [illegible]tes a chapter to him in his *Essays*. There is also an appreciation in [illegible]'s *Zur Erinnerung*. His work on the phosphoric acids is in *Ale*[illegible] *Reprint* No. 10, while Liebig's paper on the polybasic organic acid[illegible]e found in OSTWALD'S *Klassiker* No. 26.

FARADAY'S [illegible]*ental Researches in Electricity* was published in book form between 18[illegible] and 1847. Most of the work has also been reprinted in the *Klassiker*, Nos. 86, 87, 134, and 136. Their bearing upon chemistry in general, and their influence upon contemporary thought are fully discussed in OSTWALD'S *Elektrochemie*.

CHAPTER XIV

GERHARDT AND THE CHEMICAL REFORMATION—WILLIAMSON

During the period of confusion and scepticism ushered in by the collapse of dualism, it was for a time impracticable for any comprehensive theoretical system to gain a hearing. Attempts to found such systems were, of course, soon begun, and while none of these at first gained any general acceptance, one or two contained elements of truth which later proved of value. The most important of these movements was that inaugurated by Laurent and Gerhardt.

Laurent.—Auguste Laurent was born in 1807 and began his scientific studies in the *Ecole des Mines* at Paris. Having distinguished himself in chemistry he obtained at first a subordinate position at the *École Centrale des Arts et Manufactures,* where he continued his studies and received the doctorate in 1837. In the following year he became professor at Bordeaux and remained there till 1848 when his appointment as assayer of the Mint enabled him to return to Paris. He died in 1853.

Gerhardt.—Charles Frédéric Gerhardt was born in Strasburg in 1816. His first chemical studies were undertaken in Karlsruhe from 1831 to 1833 and during the following year at Leipzig. Even at this early age he showed the natural bent of his genius by attempting a classification and revision of the formulæ of the natural silicates which won commendation from Berzelius. His father now looked for the son's assistance in the manufacture of white lead, but young Gerhardt had no patience with the industrial side, and after a brief experience of army life which he liked no better, he spent a year with Liebig at Giessen where his talents and enthusiasm won him the admiration of his teacher. After one more attempt to adapt himself to the white lead industry he quarreled definitely with his father and set out for Paris almost penniless. Here his abilities gained him the notice

CHARLES GERHARDT
1816–1856

(Facing page 150)

August Wilhelm Hofmann
1818–1892

Reproduced from the "Chemical Society Memorial Lectures" by the kind permission of the Council of the Society.

of Dumas and other scientists, but his radical theories and uncompromising way of stating them made him a thorn in the side of the more conservative. In 1841 Gerhardt obtained a professorship at Montpellier which he left in 1848 in order to work with Laurent at Paris, where they founded a school of chemistry which was destined to prove a disappointment. In 1855 Gerhardt became professor in Strasburg and died there in the following year, when fortune was just beginning to smile upon him.

There is a story to the effect that when a fellow-student at Giessen once asked Gerhardt about a big manuscript he was carrying, the latter replied that it was "The Chemistry of the Future." The story might equally well have fitted Laurent. Both were radical reformers by nature, and both had to suffer as such reformers must in the appreciation of their scientific contemporaries as well as in their personal ambitions. This drew them together, and after 1843 they did practically all their work in collaboration, hence it is practically impossible to assign to each his share in their mutual services to science. They both possessed in an unusual degree the power of discovering important relationships underlying masses of apparently unrelated facts, and as their facilities for experimentation were meagre they not infrequently placed emphasis upon experimental data which later proved to be untrustworthy. This too often permitted their opponents to discredit their conclusions without answering their arguments.

Laurent had been interested in substitution ever since the earliest work of Dumas on that subject, and had gone even farther than the latter in applying the principle. As early as 1836 he attempted a classification of organic compounds in which he considered them as substitution products of certain fundamental complexes which he called "radicals," somewhat as we now derive the aliphatic compounds from the hydrocarbons of the methane series. This system came to be spoken of as the Nucleus Theory. It had many good points but it was crushed by the denunciations of Liebig and Berzelius and so never came into general use. Leopold Gmelin, however, made it the basis of classification for his celebrated *Handbuch*, a purpose for which it was admirably adapted.

Chemical Notation in 1840.—In 1840 system was sadly lacking in organic chemistry. Compounds were still classified in the text-books according to their natural sources, and nomenclature and notation were in the worst confusion. We have seen how the dualistic theory had given to acetic acid the formula which we now assign to the anhydride, $C_4H_6O_3$, and how alcohol and ether had thus naturally become $C_4H_{10}O,H_2O$ and $C_4H_{10}O$, respectively. Liebig also had doubled the formula of tartaric acid in order to account for its basicity, and in this way most of the organic compounds had come to be formulated on what was known as a four-volume basis. This meant that one formula weight occupied in the gaseous condition the same volume as four units of hydrogen. Thanks to Dumas the vapor densities of most common volatile compounds were now well known, but on account of such anomalous cases as ammonium chloride, phosphorus pentachloride, mercury and sulphur no authoritative significance was attached to them, and no one was disturbed by the prevalent inconsistency which wrote alcohol and acetic acid on the four-volume basis while ether, hydrogen sulphide, water and carbon dioxide received two-volume formulæ. Most chemists, however, followed Berzelius in writing hydrochloric acid H_2Cl_2. For him the molecular magnitude of an acid was the amount which unites with one molecule of potassium oxide or of silver oxide and he had decided to write these KO and AgO respectively. This had the further disadvantage that it concealed the dibasic character of sulphuric acid.

The foregoing, however, accounts for only a part of the prevalent confusion. Alongside of the atomic weights of Berzelius, Wollaston's equivalents had maintained their ground and as the former declined in authority the latter came more and more into use. Maximum simplicity of formulation was, however, the only criterion for the selection of equivalents, and hence each individual felt free to choose as he pleased. Most followed Wollaston in writing C = 6, O = 8, but many made other combinations and the journal literature of the time is extremely difficult reading in consequence. Berzelius made matters worse by an ill-judged compromise. He wrote what were called barred formulæ, a line drawn through a symbol standing for a double atom. H̶O for example could be interpreted according to the reader's predilections to mean either that *one equivalent* of oxygen united with one of hydrogen to form water, or that *one volume* of oxygen united with *two volumes* of hydrogen.

Gerhardt's Atomic Weights.—Gerhardt attempted a reform by reducing all formulæ to a common volume basis. Up to 1842 he made four volumes the standard and incidentally framed an ingenious argument against the dualistic theory as applied to acids. The substance we know as acetic acid was then written $C_4H_6O_3,H_2O$ as a compound of the true hypothetical acetic acid with water. On a four-volume basis, however, water is H_4O_2 and in consequence acetic acid even as written above cannot contain water apart from the radical, and the dualistic formulation is inadmissible. A little later Laurent and Gerhardt after collating the formulæ of all organic compounds whose composition could be considered well established, found that these were all divisible by two, and they then adopted the two-volume standard. This led to a series of atomic weights essentially in accordance with Avogadro's hypothesis. About the same time Regnault was obtaining valuable results in a study of specific heats which might have been utilized to support these views, but Gerhardt was fundamentally an organic chemist and physical constants really interested him little. What he desired was a standard of chemical comparison and he sought this in a comprehensive study of a multitude of chemical reactions. How far he really was from our modern point of view is shown by the concluding words of his famous paper of 1842, "Atoms, volumes and equivalents are synonymous terms." Even here, however, he was as well off as any of his contemporaries.

Atoms and Equivalents.—The word *equivalent* had been a stumbling-block for a generation. We have seen how profitably Richter had used the idea quite in the modern sense as the weight of one element which may replace or represent another in a given chemical reaction. Wollaston, however, had introduced the *word*, and had used it to denote a fixed quantity which we might define as the "simplest practicable combining weight." The two ideas have so much in common that the word was constantly used in both senses and there arose a confused idea that atoms must be equivalent, and that an atom of base must just neutralize one of acid. This doubtless fostered the prejudice which considered all acids monobasic until the work of Graham and Liebig. Even this, however, had not been pushed to its logical conclusion and the first clear distinction between atom and equivalent was drawn by Laurent in a memorable paper published in 1846. Here he pointed out that equivalency is a relationship depending upon the nature of the reaction concerned, while the standard of molecular magnitude must be sought in the vapor density. This carried with it the conclusion that the molecules of gases like hydrogen, oxygen and nitrogen must contain two atoms. He called such gases dyadides. However simple and logical this seems to us, it made no impression at all upon his contemporaries. The times were in a state of strange confusion which Ladenburg, in commenting upon Gerhardt's original suggestions, has characterized so well that the paragraph is worth quoting entire.

"It must appear strange to any unprejudiced person that the 'equivalents' which Gerhardt proposes for the elements are the same, with a few exceptions, as the atomic weights suggested by Berzelius in 1826. It is also noteworthy that Gerhardt does not mention Berzelius or even appear to know that he is adopting the latter's figures, while Berzelius on his side evidently does not notice the agreement for he attacks Gerhardt's paper violently. What I find most remarkable of all, however, is the fact that when Gerhardt made his proposal many eminent chemists (I mention only Liebig and his pupils) were already using the very ratios of atomic weights (at least for the most important elements) which Gerhardt now recommended as new, while a few years later Gmelin's equivalents, against which Gerhardt's attack was directed, had come into general use."

To Gerhardt the idea of diatomic gases was welcome for it enabled him to treat the substitution of hydrogen by chlorine gas, for example, as a metathesis and thus bring it into line with other organic reactions. In fact he developed this idea into a comprehensive theory of chemical combination. When two substances react, according to Gerhardt, the essential thing is the formation of some simple inorganic compound while the remaining atomic groups combine with each other as they may:

$$C_6H_6 + HONO_2 = \underline{H_2O} + C_6H_5-NO_2$$

The Theory of Residues.—This is the basis of Gerhardt's Theory of Residues, sometimes spoken of as the Second

Radical Theory because the residues as in the above instance frequently happened to have the same formulae as the radicals of the dualists. Gerhardt, however, stoutly denied their identity. To his mind they differed from the old radicals fundamentally in that no electrical character was assigned to them nor any separate existence in the molecule, and no pretence was made that they could ever be isolated. In spite of the fact that Gerhardt's work was destined to lead directly to our modern structural formulæ it was an article of faith with him that true structure could never be determined. He intended his formulæ to suggest reactions of formation or decomposition and held that one and the same substance might properly be assigned different formulæ according to the relationship which it was desired to emphasize; barium sulphate, for example, according to three independent methods of formation, might with equal propriety be written BaO,SO_3; BaS,O_4 or BaO_2,SO_2, and it had been a fundamental fault of the Berzelian system that it exalted the first at the expense of the others.

The Basicity of Acids.—Gerhardt gave a special name to products of metathesis like nitrobenzene. He called them "conjugate compounds" (a word which Berzelius had already employed in another sense) and noted that when one of the reacting substances is acidic its basicity is diminished by one in consequence of the operation. He applied this rule to the formation of ethyl sulphuric acid from alcohol and sulphuric acid and used it as an argument to establish the dibasic character of the latter. He also recognized that the formation of acid salts could not be considered conclusive evidence that an acid was polybasic because the salt might add a molecule of the free acid. He attached more weight to the fact that such acids can form esters, amides and so forth which are still acidic, and he contributed new facts in support of this view.

System of Classification.—No object was dearer to Gerhardt than the attainment of a consistent rational classification for organic compounds, and he attempted this by arrangements in series, distinguishing these as *homologous*, *isologous* and *heterologous*. The idea of homology had been introduced by Schiel and used by Dumas. By isologous compounds Gerhardt understood substances of analogous function like acetic and benzoic acids whose formulæ showed some other difference than CH_2. Finally by heterologous compounds he meant substances of different function but connected by genetic relationships such as alcohol and acetic acid. These frequently though not always contained the same number of carbon atoms. According to Gerhardt every organic compound should find a place in at least two of these series, and he held that when its position in these series is fixed its whole chemical character is thereby determined; exactly as, to use his own illustration, the value of a playing card is determined by its suit and spot number.

The Amines.—The above sets forth in outline what had been accomplished up to the year 1848 (the year of Berzelius's death) when support came to the new movement from unexpected quarters. In that year Wurtz discovered the primary aliphatic amines and called attention to their remarkable resemblance to ammonia. Opinion was at first divided as to their constitution, but in the following year Hofmann prepared not only primary, but also secondary and tertiary bases by the action of ammonia upon the alkyl halides. This convinced practically everyone that the new compounds were substitution products of ammonia, or according to the expression which now became common, that they belonged to the *ammonia type.*

Williamson's Work on Ethers.—In 1850 Williamson began his work upon the ethers which was destined to furnish Gerhardt with the "*terme de comparaison*" he had so long been seeking. Williamson was attempting to prepare new compounds. As Hofmann by treating ammonia with alkyl halides had obtained substituted ammonias Williamson hoped to prepare a substituted alcohol by treating potassium alcoholate with ethyl iodide. Instead he obtained ordinary ether which at first surprised him, but he was quick to see that the experiment had furnished the key to many a puzzling problem in chemistry. At this time most chemists were writing alcohol C_4H_5O,HO, the ethylate C_4H_5O,KO, and ether C_4H_5O. ($C=6$, $O=8$.) Laurent and Gerhardt, however, guided by the vapor densities, were already writing them as substitution products of water:

$$\left.\begin{matrix} C_2H_5 \\ H \end{matrix}\right\} O, \quad \left.\begin{matrix} C_2H_5 \\ K \end{matrix}\right\} O, \quad \left.\begin{matrix} C_2H_5 \\ C_2H_5 \end{matrix}\right\} O$$

Williamson saw at once that this latter view harmonized especially well with the results of his experiment, which he now formulated:

$$\left.\begin{matrix} C_2H_5 \\ K \end{matrix}\right\} O + C_2H_5I = KI + \left.\begin{matrix} C_2H_5 \\ C_2H_5 \end{matrix}\right\} O$$

It was still possible, however, to interpret the reaction on the old basis if one assumed that the ethylate first split into potassium oxide and ether, while the oxide then reacted with the iodide to form a second molecule of ether:

$$\begin{aligned} &\text{I. } C_4H_5O,KO &= KO + C_4H_5O \\ &\text{II. } C_4H_5I + KO &= KI + C_4H_5O \end{aligned}$$

Williamson disposed of this possibility in an extremely elegant and simple way by treating potassium ethylate with *methyl* iodide. If the above interpretation were correct the reaction should produce equivalent amounts of ethyl and methyl ether:

$$\begin{aligned} &\text{I. } C_4H_5O,KO &= KO + C_4H_5O \\ &\text{II. } C_2H_3I + KO &= KI + C_2H_3O \end{aligned}$$

If, on the other hand, Laurent and Gerhardt's view of the constitution of ether was correct then a new compound, methyl ethyl ether, should be the sole organic product of the reaction:

$$\left.\begin{matrix} C_2H_5 \\ K \end{matrix}\right\} O + \left.\begin{matrix} CH_3 \\ I \end{matrix}\right\} = KI + \left.\begin{matrix} C_2H_5 \\ CH_3 \end{matrix}\right\} O$$

The experiment decided in the latter sense, and the first acceptable *proof* for the new view was thereby furnished. Laurent and Gerhardt had studied hundreds of reactions and shown that their system was rational and self-consistent, but all this had carried little conviction because some other explanation was always possible. Williamson, however, had now shown by incontestable chemical evidence that, in one case at least, the formu[illegible] correspond to the vapor density. This gave the ne[illegible] an experimental foundation and it now began to ma[illegible]

The Water Type.—Williamson himself followed up these experiments first by the study of other ethers, and then by showing that not only ethers but also alcohols, esters and acids belong to the *water type.* He also expanded this idea to include polybasic acids which he considered as derived from two or more molecules of water, writing sulphuric and phosphoric acids, for example, as

$$\left.\begin{matrix} SO_2 \\ H_2 \end{matrix}\right\} O_2 \quad \text{and} \quad \left.\begin{matrix} PO \\ H_3 \end{matrix}\right\} O_3$$

This idea was still further developed when Berthelot in 1854 showed that glycerol stood in the same relation to alcohol as phosphoric acid to nitric. This led Wurtz to the discovery of glycol, the simplest diatomic alcohol, and he formulated the two compounds:

$$\left.\begin{matrix} C_6H_5 \\ H_3 \end{matrix}\right\} O_6 \quad \text{and} \quad \left.\begin{matrix} C_4H_4 \\ H_2 \end{matrix}\right\} O_4 \quad (C = 6, O = 8)$$

The Type Theory.—Williamson's discoveries gave fresh inspiration to Gerhardt, who soon followed his example by treating salts with acyl chlorides and so obtained the anhydrides of the monobasic acids.

$$\left.\begin{matrix} C_2H_3O \\ K \end{matrix}\right\} O + C_2H_3OCl = \left.\begin{matrix} C_2H_3O \\ C_2H_3O \end{matrix}\right\} O + KCl$$

This was perhaps his most appreciated experimental work. As we have seen, however, his chief talent was the organization of systems, and to the two types of water and ammonia he now added hydrogen HH, and hydrochloric acid HCl, deriving from the former the hydrocarbons and metal alkyls and from the latter the alkyl and acyl halides and the salts of the organic bases.

$$\left\{\begin{matrix} HC_2H_4 \\ HC_2H_4 \end{matrix}\right. \qquad \left\{\begin{matrix} C_2H_5 \\ Cl \end{matrix}\right. \qquad \left\{\begin{matrix} C_2H_3O \\ Cl \end{matrix}\right.$$

Butane — Ethyl chloride — Acetyl chloride

Thus originated the second Type Theory, a system far more definite and comprehensive than that of Dumas, for Gerhardt

Charles Adolfe Wurtz
1817–1894

ed into it all organic compounds. It won friends also among conservatives, for the groups which these found substituted he types were the old radicals, and Liebig who had been a ent opponent of many of Gerhardt's views had some good ds for this feature of his system. Gerhardt, however, now always, denied the radicals any objective reality, and conred the types themselves only as empty forms suitable for rpreting reactions and for classification. He used them for purpose in his *Traité de Chimie Organique*, the work by which is still best known. There is something pathetic in the fact t throughout the descriptive portion of this work he felt ged to use the old formulæ and atomic weights which he had nt his life in combating,[1] so little confidence did he feel that own arguments had made enough impression to be underod. The book itself, however, enjoyed the fullest recognition l converted many though by no means all to the new views. fortunately Gerhardt died just as it was completed and so sed even this partial triumph.

Williamson.—As we have seen, it was the work of Williamson, ırtz and Hofmann which most effectually seconded the efforts Gerhardt in establishing the Type Theory. Of these Alexander illiam Williamson was born in London in 1824. He studied st with Leopold Gmelin and then with Liebig at Giessen. ıally he devoted three years to the study of mathematics h Comte in Paris. In 1849 he became professor at the Unisity College in London and remained connected with that titution throughout his active life. His important researches e carried out between 1850 and 1860 and for the most part e connected with the classic work upon ether which we have eady considered. He died in 1904.

Vurtz.—Charles Adolphe Wurtz was born in Wolfesheim near asburg in 1817 and was for a time a schoolmate of Gerhardt. e him he studied with Liebig and later became an assistant Dumas. Being more fortunate in his friendships, however, achieved a higher material success, succeeding Dumas at *École de Médecine* in 1853 and in 1875 becoming professor

A friend who asked him why he had not clung to his own formulæ which e so much clearer, received the laughing reply: "Then no one would e bought my book!"

at the *Sorbonne.* In addition to the work on amines and glycols already alluded to may be mentioned the well-known synthesis of hydrocarbons still associated with his name, and valuable contributions to the chemistry of the metallic hydrides, of the organic compounds of phosphorus, and of the hydroxy acids. Wurtz also contributed helpfully to the view that abnormally low vapor densities are due to dissociation. His literary activities were extensive, and among his works is an *Histoire des Doctrines Chimiques* whose opening sentences, "Chemistry is a French science. It was founded by Lavoisier of immortal memory," have proved a veritable apple of discord among chemists of extreme national susceptibility. Wurtz died in 1894.

Hofmann.—August Wilhelm Hofmann was born in Giessen in 1818. He entered the university there in 1836 with the intention of studying law. Later he came under Liebig's influence and decided to devote himself to chemistry which was destined to offer him an unusually brilliant and successful career.

Having obtained the doctorate in 1841 he continued his studies with Liebig and became his assistant in 1843. In 1845 Hofmann accepted the position of docent at Bonn, and in the same year he was called to a professorship in the newly founded Royal College of Chemistry in London. Liebig's work on agricultural chemistry had aroused great interest in England, and the founders of the new institution desired to secure as its head some one who had been closely associated with Liebig. The latter suggested Hofmann, and the Prince Consort himself took an active interest in the appointment. Hofmann proved a tireless investigator and an unusually efficient and inspiring teacher. While in England he numbered among his pupils such men as Crookes, Abel and Perkin. Hofmann's own studies had begun with aniline, and throughout his life most of his investigations bore some relation to compounds of that class. It was in his laboratory in 1856 that Perkin prepared the first aniline dye, *mauve,* and one of his assistants, Peter Griess, made the fundamental studies upon diazo compounds which have later proved of so much importance in color chemistry. In general it may be said of Hofmann's work that while seldom engaged directly with industrial problems it dealt continually with those fundamental principles of organic chemistry which form the basis of the coal-tar industry. For this reason Germans are accustomed to date the beginning of their preëminence in this particular field from Hofmann's return to his native country. This took place in 1864, a call to Bonn being rapidly followed by one to Berlin where he continued his work with undiminished vigor till his death in 1892.

Hofmann was exceptionally happy as a writer and speaker. As president of the German Chemical Society it frequently fell to his lot to deliver memorial addresses or write obituary notices which he wrought into works of real biographical value. Most of these have been published under the title *Zur Erinnerung an Vorangegangene Freunde* and constitute a veritable treasury of chemical reminiscence and appreciation.

Literature

Charles Gerhardt, sa Vie, son Oeuvre, sa Correspondance, by his son CHARLES GERHARDT and E. GRIMAUX, Paris, 1900, is the authoritative biography. It makes, however, a painful impression by constantly holding up Gerhardt as a martyr at the hands of his contemporaries. There is also an interesting study of Gerhardt in OSTWALD'S *Grosse Männer.* It depends, however, upon the above biography for its facts. *Le Centenaire de Charles Gerhardt* published by the Société Chimique de France in 1916, contains a sympathetic account of his work by Marc. Tiffenau. GERHARDT'S own *Traité de Chimie Organique* is still interesting reading.

Williamson's work on ether was reprinted by the Alembic Club, No. 16. That of Wurtz on ethylene glycol is to be found in the *Klassiker* No. 170. There is an especially interesting memorial to Wurtz in HOFMANN'S *Zur Erinnerung;* also another by FRIEDEL in the *Bulletin de la Société Chimique* for 1888.

The *Chemical Society Memorial Lectures,* London, 1901, contain appreciations of Hofmann by PLAYFAIR, ABEL, PERKIN, and ARMSTRONG. The German Chemical Society also published a memorial volume in 1900 entitled *August Wilhelm von Hofmann, ein Lebensbild.* The biographical portion is written by JACOB VOLHARD, the chemical by EMIL FISCHER.

CHAPTER XV

THE TRANSITION FROM THE TYPE THEORY TO THE VALENCE THEORY

While Laurent and Gerhardt were laying the foundations of the Type Theory, two other chemists, Kolbe and Frankland, were following an entirely different line of thought.

Kolbe.—Adolph Wilhelm Hermann Kolbe was born at Elliehausen near Göttingen in 1818, and at the age of twenty began the study of chemistry with Wöhler. In 1842 he became an assistant of Bunsen at Marburg, and three years later went to London where he worked under Playfair. From 1847 till 1851 he devoted himself to literary work in connection with various chemical publications. In the latter year he succeeded Bunsen at Marburg, and in 1865 became professor at Leipzig where he remained till his death in 1884. Kolbe was eminent as a teacher nd a highly original thinker, distinguished alike for the brilliancy of his ideas and the caustic virulence of his polemical writings.

Frankland.—Edward Frankland was born at Churchtown near Lancaster in 1825. After six years as an apothecary's assistant he began the systematic study of chemistry under Playfair in London. Here he met Kolbe and then like him worked for a time with Bunsen. Returning to England in 1847 he at first accepted a position in a school till in 1851 he obtained a professorship in Owens College, Manchester. Six years afterward he became lecturer at St. Bartholomew's Hospital and in 1863 professor at the Royal Institution. Two years later he succeeded Hofmann at the School of Mines with which the Royal College of Chemistry had just been merged. Frankland died on a visit to Norway in 1899. In addition to the work which we are about to consider he made many noteworthy contributions to organic chemistry and did practical work of much value in connection with the London water supply. In 1868 while engaged in a study of the solar spectrum in collaboration with Sir Norman Lockyer they observed certain lines which could not be attributed to any element then known to exist on the earth, and to this unknown substance they gave the name *helium.*

Frankland as well as Kolbe found the views of Gerhardt unsympathetic, and his types too artificial to be accepted as the foundation of a true chemical system. They therefore endeavored to construct something better by turning to account what was good in the old radical theory.

Conjugate Compounds.—It will be recalled (page 136) that this theory had been shipwrecked by Dumas's discovery of trichloroacetic acid and its close chemical resemblance to the parent substance. Berzelius was then writing 'anhydrous' acetic acid $C_4H_3O_3$ (C=6, O=8) as a compound of negative oxygen with the positive radical C_4H_3. He could not, however, accept for the chlorine compound the corresponding formula $C_4Cl_3O_3$ because this would involve the incorporation of a negative element in the radical. The best he could do was to modify the formula of acetic acid and write it C_2H_3,C_2O_3 as a 'conjugate' compound (addition product—see also page 155) of 'anhydrous' oxalic acid and 'methyl' which he called the *copula.* On chlorination, the oxalic acid, a true electrochemical compound, remained unchanged, while the copula (which had really been invented for this purpose) became substituted. This was generally regarded as a makeshift which really conceded all that it had been invented to avoid. Berzelius, however, never admitted this, and Kolbe and Frankland felt that if they could show that acetic acid really contained methyl and oxalic acid the theory would appear in a better light.

The Kolbe Synthesis.—Kolbe believed he had done this when he carried out the electrolysis of acetic acid. This reaction, which has since become an important synthesis, we now interpret as follows: Acetic acid CH_3CO_2H dissociates into the ions H and CH_3CO_2. On electrolysis the hydrogen appears at the cathode, while at the anode the other ion CH_3CO_2 breaks up, forming carbon dioxide and ethane, the latter being produced by the combination of two methyl groups which the evolution of carbon dioxide has set free:

$$\begin{matrix} CH_3CO_2H \\ CH_3CO_2H \end{matrix} = H_2 + 2\ CO_2 + \begin{matrix} CH_3 \\ | \\ CH_3 \end{matrix}$$

Kolbe's explanation seemed more simple. For him acetic acid is methyl plus oxalic acid, and on electrolysis the latter is oxidized to carbonic acid while the methyl is set free:

$$C_2H_3,C_2O_3 + O = C_2H_3 + 2CO_2$$

We now know of course that what Kolbe really obtained was not methyl but ethane. The proof of this, however, could not be furnished until some years later, so that Kolbe's conclusion seemed exceedingly plausible.

Other reactions were also observed which it was found possible to interpret in the same sense. Frankland and Kolbe were the first to carry out the hydrolysis of acetonitrile which we now formulate:

$$CH_3.CN + 2H_2O = CH_3.COOH + NH_3.$$

Like us they regarded the nitrile as a compound of methyl with cyanogen, C_2H_3,C_2N, and since the latter is known to yield oxalic acid on hydrolysis, they regarded the formation of acetic acid as a confirmation of their view:

$$C_2H_3,C_2N + 3HO = C_2H_3,C_2O_3 + NH_3.$$

The Metal Alkyls.—Finally in 1849, Frankland, by treating ethyl iodide with zinc, obtained a substance which we now know to be butane but which he considered as free ethyl:

$$C_4H_5I + Zn = ZnI + C_4H_5$$

just as Kolbe had considered the product of his electrolysis to be free methyl. Since this passed undisputed at the time, Frankland considered it as the fullest justification of his views, for to the believers in the old radical theory the best evidence for the existence of a radical was its isolation.

As by-products in the last reaction Frankland observed the formation of the zinc alkyls:

$$C_4H_5I + 2Zn = ZnI + ZnC_4H_5$$

substances which aroused great interest on account of their unexpected composition, their remarkable physical properties, and the synthetic reactions which could be accomplished by their use. It is perhaps upon this discovery that the fame of Frankland chiefly rests. The further study of these compounds, however, caused him to abandon his electrochemical views. In accordance with the latter the copula should have practically no influence upon the combining capacity of the other constituents of a compound, and zinc methyl, for example, ought when oxidized to yield zinc oxide methyl:

$$ZnC_2H_3 + O = ZnO,C_2H_3$$

No such reaction could, however, be carried out, and Frankland found it perfectly general that the power of a metal to unite with oxygen was diminished by one[1] for every alkyl group with which it was combined. Thus cacodyl oxide $As\left\{\begin{array}{l}C_2H_3\\C_2H_3\\O\end{array}\right.$ could only be oxidized to $As\left\{\begin{array}{l}C_2H_3\\C_2H_3\\O\\O\\O\end{array}\right.$, and stibethine $Sb\left\{\begin{array}{l}C_4H_5\\C_4H_5\\C_4H_5\end{array}\right.$ to $Sb\left\{\begin{array}{l}C_4H_5\\C_4H_5\\C_4H_5\\O\\O\end{array}\right.$. In short, cacodyl oxide and stibethine are arsenious oxide and antimony oxide respectively in which more or less oxygen has been replaced by radicals. This is, however, just the point of view of the type theory, and though the types are not exactly those of Laurent and Gerhardt, Frankland now agrees essentially with them in regarding organic compounds as inorganic ones in which some element has been *substituted* by a radical. In the course of his general discussion Frankland writes as follows:

Frankland on Valence.—"When the formulæ of inorganic chemical compounds are considered, even a superficial observer is struck with the general symmetry of their constitution; the compounds of nitrogen, phosphorus, antimony and arsenic especially exhibit the tendency of these elements to form compounds containing 3 or 5 equivalents of other elements, and it is in these proportions that their affinities are best satisfied; thus in the ternal group we have NO_3, NH_3, NI_3, NS_3,

[1] *One,* that is, on the basis of the atomic weights which he was using, C = 6, O = 8, etc.

PO_3, PH_3, PCl_3, SbO_3, SbH_3, $SbCl_3$, AsO_3, AsH_3, $AsCl_3$, etc., and in the five atom group NO_5, NH_4O, NH_4I, PO_5, PH_4I, etc. Without offering any hypothesis regarding the cause of this symmetrical grouping of atoms, it is sufficiently evident from the examples just given, that such a tendency or law prevails, and that, no matter what the character of the uniting atoms may be, the combining power of the attracting element, if I may be allowed the term, is always satisfied by the same number of these atoms."

This passage, which was written in 1852, is often spoken of as the first statement of the valence theory, and in a sense this is true, at least so far as the principle is concerned. The *idea* of valence, however, could be of little value until the principles were established by which the actual valence of a given element could be determined, and here there was as yet no uniformity of opinion. Indeed the most superficial examination of the above list of formulæ shows how little prepared Frankland himself was at this time to determine the actual valence even of oxygen. It is therefore quite natural that the idea bore no fruit until some years later.

Kolbe's Notation.—Kolbe was more conservative than Frankland and therefore slower to surrender his electrochemical ideas. Finally, however, he came to adopt the view that organic compounds are best considered as substitution products of inorganic ones, and he worked out a complex and highly original system in which he formulated practically all organic compounds as substitution products of carbonic acid. There is no occasion to discuss the details of this system, for it was hardly employed save by Kolbe and those under his immediate influence, but by its aid he was able to give account of many reactions more clearly than Wurtz and Gerhardt, and sometimes to predict compounds and reactions as yet unknown. The case most often quoted is that of the secondary and tertiary alcohols of which he not only predicted the discovery but also the behavior on oxidation. This was rightly hailed as a great triumph, and might have brought his system into general recognition had not the valence theory soon after supervened.

In general it may be said that the chief service of Frankland and Kolbe to chemical theory was the emphasis they placed upon the reality of the radicals. This compelled the disciples of

Gerhardt to regard their own types as an expression of chemical constitution, and deterred them in some measure from classifying unlike things together on the sole basis of some purely formal analogy in their composition.

We have next to trace the steps by which the type theory grew into the valence theory. One of the greatest leaders in this movement was Friedrich August Kekulé.

Kekulé.—Kekulé was born in Darmstadt in 1829. He showed early talent in drawing and entered the university of Giessen with the intention of studying architecture. Here, however, Liebig's influence decided him for chemistry. After some years in Paris, where he came in contact with Gerhardt and other celebrated French scientists, Kekulé returned to Germany in 1856 as a docent at Heidelberg. Here he did some of his most important work, and in 1858 he was called to a professorship at Ghent. Ten years later he was made professor at Bonn, where he remained till his death in 1896. Kekulé always distinguished himself as a brilliant and daring thinker especially devoted to organic chemistry, a branch of the science which is indebted to his inspiration for many of its fundamental assumptions.

Multiple Types.—The first step toward a logical expansion of the type theory had been taken by Williamson when he introduced the so-called "multiple types." We have seen an example of these in the derivation of sulphuric acid from two molecules of water (page 158). Written out in full this took the form:

$$\begin{array}{rl} \mathrm{H} & \\ & \}\,\mathrm{O} \\ \mathrm{SO_2} & \{ \\ & \}\,\mathrm{O} \\ \mathrm{H} & \end{array}$$

and Williamson pointed out that the diatomic group SO_2 really took the place of one hydrogen in each of two molecules of water, thus holding them together:

$$\left.\begin{array}{r}\mathrm{H}\\ \mathrm{H}\end{array}\right\}\mathrm{O}$$
$$\left.\begin{array}{r}\mathrm{H}\\ \mathrm{H}\end{array}\right\}\mathrm{O}$$

A similar notation was quite extensively employed by others, and Kekulé in 1854 applied the idea in a somewhat unusual way to the action of phosphorus pentasulphide upon acetic acid forming thioacetic acid. He showed that this is really parallel to that of phosphorus pentachloride, oxygen in one case being replaced by sulphur and in the other by chlorine. In the latter case, however, "the product decomposes into acetyl chloride and hydrochloric acid, whereas when the sulphur compound of phosphorus is employed the groups remain together *because the quantity of sulphur which is equivalent to two atoms of chlorine is not divisible.*" He writes:

$$5\left.\begin{matrix}C_2H_3O\\ H\end{matrix}\right\}O + P_2S_5 = 5\left.\begin{matrix}C_2H_3O\\ H\end{matrix}\right\}S + P_2O_5$$

$$5\left.\begin{matrix}C_2H_3O\\ H\end{matrix}\right\}O + 2PCl_5 = \frac{5C_2H_3O.Cl}{5HCl} + P_2O_5\ .$$

and cites as another example the action of the same reagents upon alcohol:

$$\left.\begin{matrix}C_2H_5\\ H\end{matrix}\right\}O \xrightarrow{P_2S_5} \left.\begin{matrix}C_2H_5\\ H\end{matrix}\right\}S \quad \text{but} \quad \left.\begin{matrix}C_2H_5\\ H\end{matrix}\right\}O \xrightarrow{PCl_5} \frac{C_2H_5.Cl}{HCl}$$

This of course only emphasizes in another way what Williamson had shown when he established the fact that oxygen can unite with two dissimilar radicals. Kekulé goes on to say:

"It is not only a difference in formulation but an actual fact, that one atom of water contains two atoms of hydrogen and one atom of oxygen, and that the quantity of chlorine equivalent to one indivisible atom of oxygen is itself divisible by two, whereas sulphur like oxygen is dibasic so that one atom of sulphur is equivalent to two of chlorine."

Mixed Types.—A further advance was made in 1857 when Kekulé revived the marsh-gas type of Dumas and added it to hydrogen, water and ammonia. About the same time he introduced the idea of "mixed types." These resembled the multiple types but included the simultaneous use of molecules of different types. In this way it was possible to derive, for example,

FRIEDRICH AUGUST KEKULÉ
1829–1896

Reproduced from the "Chemical Society Memorial Lectures" by the kind permission of the Council of the Society.

(Facing page 170)

H }
{H }
{H }
) from }O and
{H }
{H }
}O
H }

H }
H }N
SO₂ }
}O
H }

Sulphamic acid

from
H }
H }N
{H }
{H }
}O
H }

ıu-

·hand brackets in the type formula designating in each
: hydrogen atoms which are replaced by a polyatomic
It is interesting in this connection to see how Kekulé
a radical:

rding to our view the radicals are nothing but the residues
tacked by a given decomposition. In one and the same sub-
herefore, according as a greater or smaller part of the atomic
is attacked, we may assume a greater or smaller radical.
example, when we consider the salt formation of sulphuric
are led to the conclusion that it contains the radical SO_4. It
then as water in which oxygen is replaced by the radical SO_4
mparable with hydrogen sulphide.

$$\left.\begin{matrix}H\\H\end{matrix}\right\}O \qquad \left.\begin{matrix}H\\H\end{matrix}\right\}S \qquad \left.\begin{matrix}H\\H\end{matrix}\right\}SO_4$$

owever, we consider the action of phosphorus pentachloride
:hat the group SO_4 contains two atoms of oxygen which are
ɔle by chlorine; we have

$$\left.\begin{matrix}H\\SO_2\end{matrix}\right\}O \; \left.\begin{matrix}\\H\end{matrix}\right\}O \longrightarrow \frac{\frac{HO}{SO_2Cl}}{HCl} \longrightarrow \frac{\frac{HCl}{SO_2.Cl_2}}{HCl}$$

nust therefore assume that the radical SO_2 is present in sul-
:id. A more penetrating decomposition shows us therefore
group which in other decompositions remains unchanged
as a radical) is really only a compound of another radical
tion of the radicals.)"

last phrase, the constitution of the radicals.

shows the half-conscious tendency of the times toward explaining the transformations of the radicals by the arrangement of their component atoms. Kekulé himself did not go as far as this for some time, but he shows how interesting relationships can be emphasized by writing compounds as derivatives of different types, dinitrobenzene, for example, and phenylene diamine can have either of the following formulæ:

$$\left\{\begin{matrix} H \\ \\ C_6H_3(NO_2)_2 \end{matrix}\right. \text{ or } \left\{\begin{matrix} C_6H_4 \\ \\ (NO_2)_2 \end{matrix}\right. \quad \left\{\begin{matrix} H \\ \\ C_6H_3(NH_2)_2 \end{matrix}\right. \text{ or } N\left\{\begin{matrix} H \\ H \\ C_6H_4.NH_2 \end{matrix}\right.$$

It may be remarked in passing that the theory of mixed types led to some formulæ well-nigh grotesque in their complexity, and that Kolbe held the whole idea up to ridicule on this ground. The criticism would, however, have probably made a deeper impression had it come from some other than Kolbe, whose own formulæ were not always of the simplest. The following, for example, shows the way in which he wrote sulphoacetic acid:

$$2HO\left\{ C_2\left\{\begin{matrix} H \\ SO_2 \\ SO_3 \end{matrix}\right\}\frown C_2,O_3 \right\}$$

In 1858 Kekulé published a celebrated paper in which he further emphasized the importance of the methane type and showed that whenever a hydrogen in methane is replaced, the carbon and the remaining hydrogen constitute a radical whose valence is increased by one:

$$CH_4 \longrightarrow CH_3Cl \longrightarrow CH_2Cl_2 \longrightarrow CHCl_3 \longrightarrow CCl_4$$

methyl being monatomic, methylene diatomic, and so on. He then took up the numerical ratio of hydrogen to carbon in the radicals, C_nH_{2n+1}, and the hydrocarbons, and came to the conclusion that when several carbon atoms occur together in a compound radical *they are connected with each other*. This idea, which now seems axiomatic, was highly original. It was scarcely implied in the older theories and was indeed rather foreign to their point of view. Once grasped, however, it gave the key to the constitution of all organic compounds. Kekulé himself, nevertheless, did not at once begin to write graphic formulæ, but rather expressed himself with great conservatism in words which might have been written by Gerhardt:

"Rational formulæ are transformation formulæ (*Umsetzungsformeln*) and in the present state of the science they can be nothing else. By showing on the one hand the atomic groups which remain unaffected by certain reactions (the radicals), and on the other those which play a rôle in frequently recurring metamorphoses (the types), such formulæ give a picture of the chemical nature of the substance. Every formula which shows some reactions of a compound is *rational*. Of the different rational formulæ that which at the same time expresses the greatest number of metamorphoses is the *most rational*."

First Graphic Formulæ.—The complete analysis of organic radicals down to the arrangement of their component atoms was first attempted by Couper in the same year, and independently of Kekulé. Couper wrote graphic formulæ in the modern sense. He saw fit, indeed, to halve the atomic weight of oxygen, but since he assumed that two atoms of this element always occurred together in any organic compound his formulæ are essentially like our own. The following examples will suffice:

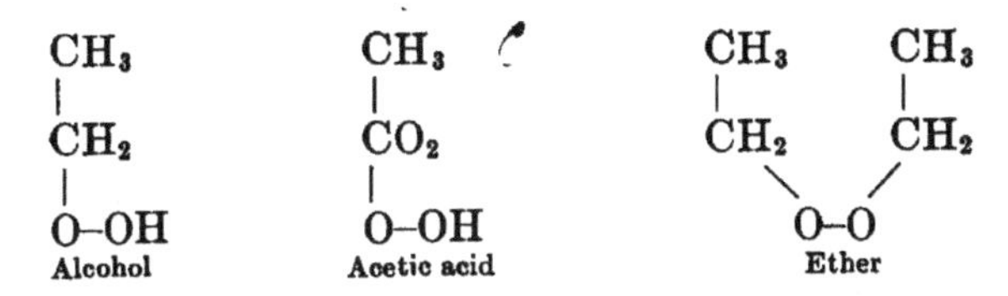

Alcohol Acetic acid Ether

In 1861 appeared the first portion of Kekulé's great text-book which emphasized and illustrated the new views with hundreds of examples. The foundations of modern organic chemistry were therein laid and, what is more important for us here, the date marks the time when the great contribution of organic chemistry to the historical development of the science as a whole was fully rendered. The theory of electrochemical dualism had broken down because it had failed to explain the reactions of organic chemistry. Slowly and laboriously through the stages of the type theory there had grown up in organic chemistry the unitary theory of structure which was now destined to become dominant in its turn.

The Vindication of Avogadro's Hypothesis.—The real service of organic chemistry had, however, been greater. It completed the atomic theory. Dalton had no sooner put forward the fundamental idea than he was confronted with the question; what are the atomic weights? This was seen to depend upon the composition of simple compounds. What is the formula for water? for ammonia? for methane? Dalton could not answer these questions, nor could any chemist answer them satisfactorily for fifty years. Avogadro's hypothesis seemed to offer hope of a solution but it had been disregarded because for a long time it seemed only applicable to the few cases it had been designed to explain. The laws of Mitscherlich and of Dulong and Petit had promised much, and yet had proved in practice equally inconclusive. It was reserved for the typists to find more convincing arguments. Williamson's preparation of methyl ethyl ether first proved by chemical means that water could not have a simpler formula than H_2O, the work of Hofmann and Wurtz on the amines showed that the formula of ammonia could not be simpler than NH_3, and that of Kekulé and others on many organic substances fixed that of methane as CH_4. When now the vapor densities of all these compounds were studied they proved throughout to have two volume formulæ, and Avogadro's hypothesis was thereby rehabilitated.

Removal of the Difficulties.—Of course there still remained many contradictions and apparent inconsistencies to clear up. The chief difficulties which had always stood in the way of a general acceptance of the hypothesis had been, first, the anomalous vapor densities of certain inorganic compounds like ammonium chloride and phosphorus pentachloride; second, the vapor densities of elements like mercury and sulphur which seemed inharmonious with those of oxygen and hydrogen; and finally the disinclination which many felt toward believing that the smallest (physical) particles of the elements were themselves complex. General experience seems to show that chemical affinity is strongest between unlike elements, and that compounds are most stable when the components are dissimilar. How then could a molecule of hydrogen gas, for example, be made up of two atoms exactly alike? Such an idea was especially distasteful to the disciples of Berzelius, because his theory ascribed [illegible] chemical affinity to electric charges, and how could two atoms having exactly the same charge unite?

With the progress of time, however, experimental facts had accumulated in support of the discredited idea. Fabre and Silbermann in 1846 found that carbon gave off more heat when burned in nitric oxide than in pure oxygen. This could hardly be explained in any other way than by saying that what produced heat in both cases was the union of carbon and oxygen. The only thing which could absorb it, however, must be the energy required to separate the oxygen from the nitrogen in the first case, and the oxygen *from itself* in the second. In other words the oxygen molecule must be compound.

The vapor density of sulphur also diminishes with rise of temperature far more rapidly than the law of Gay-Lussac requires, and this pointed to the compound nature of at least the denser form. Finally the enhanced reactivity of elements at the moment of liberation, the so-called nascent state, was hard to explain save by assuming the transitory existence of single atoms.

The abnormal vapor densities of compounds could also now be explained. It was due to *dissociation.* Ammonium chloride when sublimed decomposes into ammonia and hydrochloric acid which under ordinary circumstances recombine on cooling. If, however, the sublimation takes place in a vessel with a porous wall the lighter ammonia diffuses through this more rapidly than the hydrochloric acid, showing that at this temperature the gases are uncombined. Similarly in the case of phosphorus pentachloride, dissociation into the trichloride and chlorine could be experimentally proved.

The hypothesis of Avogadro had explained the reactions of oxygen, hydrogen, chlorine and nitrogen by the assumption that the molecule of each gas contained two atoms. To many it apparently seemed a necessary consequence that the same must hold true of every elementary gas. Avogadro himself certainly never drew any such conclusion, but the fact that the vapor density of mercury indicated a smaller number and that of sulphur a greater seemed generally to be taken as an argument against the theory. Once the other difficulties were disposed of, however, it was not difficult to show that there need be no uniformity among the elements in this respect.

The Service of Cannizzaro.—The foregoing facts were all known or at least available in 1858, but no one had summed them up or wrought them into a conclusive argument in support of the hypothesis, and the sad confusion of atomic weights and of chemical notation which had been growing worse since 1840 showed little sign of improvement. In 1860 at the instance of Weltzien, Wurtz and Kekulé a convention was called in the hope of bringing about some general understanding or at least some formal compromise. The meeting took place at Karlsruhe in September of that year, Dumas presided, and the other great lights of the science were well represented. Among those present was Stanislao Cannizzaro (1826–1910) then professor in Genoa but later in Palermo and at Rome. Two years before, he had written a little pamphlet entitled *Sunto di un Corso di Filosofia Chimica* describing the plan of instruction by which he was accustomed to introduce his own students to the subject of theoretical chemistry. He made Avogadro's hypothesis the foundation of his system, and showed with exemplary clearness and abundance of illustrative detail how this theory accounts for all forms of chemical combination and how the apparent contradictions were to be explained.

The convention proceeded as such assemblies commonly do. Many brilliant speeches were made but no general agreement was reached. Just at the close, however, the little booklet of Cannizzaro was distributed and seems to have made a wonderful impression upon all who read it. Lothar Meyer thus describes the effect upon himself:

"I also received a copy which I put in my pocket to read on the way home. Once arrived there I read it again repeatedly and was astonished at the clearness with which the little book illuminated the most important points of controversy. The scales seemed to fall from my eyes. Doubts disappeared and a feeling of quiet certainty took their place. If some years later I was myself able to contribute something toward clearing the situation and calming heated spirits no small part of the credit is due to this pamphlet of Cannizzaro. Like me it must have affected many others who attended the convention. The big waves of controversy began to subside, and more and more the old atomic weights of

Berzelius came to their own. As soon as the apparent discrepancies between Avogadro's rule and that of Dulong and Petit had been removed by Cannizzaro both were found capable of practically universal application, and so the foundation was laid for determining the valence of the elements, without which the theory of atomic linking could certainly never have been developed."

Once Avogadro's hypothesis had been definitely accepted the world possessed a reliable standard of atomic and molecular magnitude, and the investigator was able for the first time to feel sure when he was dealing with comparable quantities in the case of elements as well as of compounds. It is hardly too much to say that modern chemistry began in 1860.

Literature

The work of Kolbe upon the electrolysis of organic acids is found in *Alembic Club Reprint* No. 15, and his system for the formulation of all organic compounds as derivatives of carbonic acid in the *Klassiker* No. 92.

A memorial lecture on Frankland was delivered by ARMSTRONG before the Chemical Society of London in 1901, and printed in brief abstract in the *Proceedings* for that year, page 193.

A German translation of the pamphlet by CANNIZZARO, *Sunto di un Corso di Filosofia Chimica*, is printed entire in OSTWALD'S *Klassiker* No. 30.

THORPE'S collection contains essays on Cannizzaro and Kekulé.

KOPP'S *Theoretische Chemie* is an extremely interesting book for the student of this period. It is no easy task for the chemist of the twentieth century to 'think himself back' into the point of view of the type theory, and to the one who desires to accomplish this KOPP'S book can render valuable assistance. Though published in 1863 it contains no mention of the valence theory and discusses the relative merits of the views of Williamson and Kolbe with great clearness from a contemporary standpoint.

CHAPTER XVI

THE PERIODIC LAW

Until Williamson and his co-workers had fixed the formula of water there was no agreement as to whether the atomic weight of oxygen was eight or sixteen, and a similar uncertainty was the rule with reference to the other atomic weights. In such circumstances it was really idle to discuss numerical relationships which might exist between them. By 1860, however, data had become available for fixing these numbers, and their mutual relationships attracted more and more attention.

The Atomic Weights.—If we may define an atom as the smallest quantity of an element known to exist in any of its compounds, and an atomic weight as the relative weight of such atoms, then the determination of this value involves two distinct problems. The first is exclusively analytical and is directed toward fixing the proportion by weight in which a given element unites with others. The second is concerned with determining what multiple or submultiple of the combining weight fulfills the other condition. We have seen how Berzelius had made it the chief object of his life to answer the first of these questions, and he performed his work with such thoroughness as to earn from Dumas the splendid tribute: "Whoever works under the same conditions as Berzelius obtains the same results as Berzelius. Otherwise he has not worked correctly." New times, however, bring new conditions and with improvement in technique all scientific measurements require revision. Dumas himself did valuable work of this kind, his determination of the oxygen hydrogen ratio in water being long considered a masterpiece.

Stas.—All previous work upon atomic weights, however, was surpassed by that of the Belgian chemist, Jean Servais Stas. Stas was born in Louvain in 1813. He originally took a degree in medicine and it was his interest in substances of physiological importance which first led him to enter Dumas's laboratory. Here in addition to his organic work he coöperated with Dumas in a determination of the atomic weight of carbon. In 1840 he began to teach in the military school at Brussels where he remained nearly twenty-five years, but at last became incapacitated for teaching on account of an ailment which affected his speech. He next accepted a position in the mint which he resigned in 1872, living in retirement at Brussels till his death in 1891. Stas had early become interested in Prout's hypothesis, and resolved to devote his life to testing it by the most accurate possible determination of the combining weights. He refined the processes of quantitative transformation to a degree never before equalled, working with highly purified materials, employing exceptional weights of substance, making his weighings upon balances of hitherto unequalled precision, and exercising extraordinary care in his manipulations. Some of his precautions were, as a matter of fact, illusory, so far as their bearing upon the accuracy of the final results are concerned, as has been shown by the still more accurate figures, which have been obtained in recent years by Theodore William Richards and his co-workers in the Harvard laboratory, but none the less, the work of Stas still represents the maximum of human patience applied to quantitative analysis. His results convinced Stas that there was nothing in Prout's hypothesis unless one were willing to assume that the atomic weight of the primal substance was smaller than the experimental error in his determinations, which of course begs the entire question.

The second question: what multiple or submultiple of the combining weight is the atomic weight, was now at last in some position to be answered. It had been settled for most cases by the acceptance of Avogadro's hypothesis, since this fixed the molecular magnitude of compounds as the weight occupying in the gaseous state the same volume as two equivalents of hydrogen. Where this failed, as in the case of elements which form no volatile compounds, an intelligent application of the laws of atomic heats or of isomorphism could be depended upon to fix the true value with a high degree of certainty. Numerical relationships between the atomic weights taken as a whole could therefore now be profitably considered.

Döbereiner's Triads.—The subject had indeed aroused interest long before. Certain groups of elements like the halogens and the alkali metals show a resemblance in physical and chemical properties which is so striking as to impress the most casual observer. The first recognition that such similarity might have anything to do with atomic weight was recognized by Prout who in the first paper setting forth his famous hypothesis (page 81) ascribes the chemical resemblance of iron, cobalt, and nickel to the fact that they all have the same combining weight (28). A more comprehensive generalization was made by Döbereiner in 1839. He called attention to the fact that similar elements usually existed in groups of three which he called *triads,* and he showed that their combining weights possessed the numerical peculiarity that one was the mean of the other two. This applied to chlorine, bromine and iodine; to calcium, strontium and barium; to lithium, sodium and potassium; and to sulphur, selenium and tellurium, with an accuracy closer than a single unit. Platinum, iridium and osmium formed another triad, and also silver, lead and mercury (108; 104; 100). The reader will note how the true atomic weights of lead and mercury would spoil this relationship. When such classification failed, Döbereiner adopted incomplete triads. Phosphorus and arsenic, for example, represented such a group, as did also boron and silicon. Still other elements had to be classed in larger groups.

Pettenkofer and Dumas.—In 1850 Pettenkofer called attention to the fact that the currently accepted combining weights of similar elements frequently differed from each other by some multiple of eight, thus in the alkalies we have lithium 7, sodium 23 (7 + 16), potassium 39 (23 + 16); and among the alkaline earths, magnesium 12, calcium 20 (12 + 8), strontium 44 (20 + 24), barium 68 (44 + 24).

Dumas aroused unusual interest by a paper in defence of Prout's hypothesis which he delivered before the British Association in 1852. He conceded that in the triads the actual value of the middle term usually differed from the calculated one by a quantity well outside the experimental error of the determination, but he held that the elements are related by some law akin to that of homologous series in organic chemistry, wherein a simple formula of two or three terms accounts for the composition of any member of the series. In a series of similar elements, he argued that the general chemical character of the series is fixed by the equivalent weight of the lowest member, while the properties of the higher elements are determined by certain orderly increments in the combining weight. Thus in the nitrogen group, for example, we have

a	14	nitrogen	14
a + d	14 + 17	phosphorus	31
a + d + d′	14 + 17 + 44	arsenic	75
a + d + 2d′	14 + 17 + 88	antimony	119
a + d + 4d′	14 + 17 + 176	bismuth	207

He also pointed out certain relationships between dissimilar groups, showing that the difference of five between phosphorus and chlorine is repeated between arsenic and bromine and between antimony and iodine. These ideas were of course much elaborated with other examples in the original paper.

Gladstone, Cooke and Odling.—In 1853 J. H. Gladstone stated that the atomic weights of similar elements might be related in three ways: They might be the same, as in the cases of cobalt and nickel, or in that of palladium, rhodium and ruthenium; they might be in multiple proportion, as in the cases of the palladium group (53) the platinum group (99) and gold (197); or finally they might differ by a common increment as in the case of several instances previously cited.

In the next year Josiah P. Cooke of Harvard College made a classification more elaborate and complete than any which had preceded it, and embracing many interesting details which cannot be discussed here. He divided the elements into six series each characterized by a special numerical relationship and published a series of tables bringing out these relationships as well as various other analogies among the different groups, laying special stress upon crystalline form. Like Dumas he considered that the fundamental cause of the relationships observed must be something akin to homology.

Odling in 1857 published a classification into thirteen groups which were essentially triads, though elements were included which belonged to no triad. Later, in 1864 he pointed out that there is a marked continuity when the atomic weights are written in numerical order, and he noted that similar elements are frequently separated by 48, while intervals of 16, 40 and 44 are also commonly observed. This led him to believe that 4 might represent the unit of common difference, a surmise which has gained significance in our modern theories of atomic structure.

The Helix of de Chancourtois.—In 1863, de Chancourtois published the first of a series of papers in which he brought out more clearly than had hitherto been done the fact that there is a regular recurrence of elements with similar properties when these are arranged in the order of their atomic weights. He chose a graphic method of representation. The convex surface of a vertical cylinder was ruled with 16 equidistant lines parallel to the axis, the number 16 having been chosen because it represented the atomic weight of oxygen. A helix was then drawn starting at the base and ascending the cylinder at an angle of 45°. In this way each intersection of the helix with one of the vertical lines could represent a unit of atomic weight, and the atomic weight of every element would be represented by a point on the helix. As a matter of fact, when the arrangement is complete the resemblance of elements which stand vertically above each other upon the cylinder is very marked. We are, however, not limited to vertical relationships. Any two points upon the curve can be connected by a line and when this is produced it must generate another or secondary helix, and de Chancourtois believed that all the elements whose atomic weights fell in such a line could be shown to stand to each other in some orderly chemical relationship. These ideas were interesting, but they seem to have made little impression upon contemporaries.

Newlands's Law of Octaves.—At about the same time Newlands in England was publishing a series of papers dealing with atomic classification, in the first of which he made no great advance upon his predecessors. By 1865, however, the old equivalents which had served earlier investigators as a basis of comparison had given place to the atomic weights, and Newlands found that the simplest arrangement of these was also the most striking.

JULIUS LOTHAR MEYER
1830–1898

Reproduced from the "Chemical Society Memorial Lectures" by the kind permission of the Council of the Society.

DMITRIJ IVANOVITCH MENDELEJEFF
1834–1907

He wrote the elements in the order of their atomic weights, and for convenience numbered them just as they occurred on this list. He then compared these numbers instead of the atomic weights themselves, finding that "the numbers of analogous elements, when not consecutive, differ by 7 or a multiple of 7." Thus the ninth and sixteenth closely resemble the second (Li,Na,K). Newlands found it necessary to transpose a few elements in his list in order to make the above relationship hold, but it did not seem unreasonable that more accurate determinations of some atomic weights might justify the transpositions. He called his generalization the Law of Octaves and the story is told that when he first gave an account of it to the Chemical Society in London one prominent member asked "whether he had ever examined the elements according to the order of their initial letters," a remark which showed the amount of contemporary interest in such speculations. Nevertheless Newlands had pointed out more clearly than any previous chemist the periodic recurrence of similar properties among the elements. His system, however, showed no gaps and therefore left no room for new elements unless some entire octave should be discovered. This judicious sense of where the gaps must be, characterize the more complete systems of Lothar Meyer and Mendelejeff.

Lothar Meyer.—Julius Lothar Meyer was born in Varel, Oldenburg, in 1830. He began the study of medicine at Zurich and Würzburg, and went to Heidelberg in 1854 where the influence of Bunsen and Kirchhoff drew him so far to the other extreme that we find him not long after at Königsberg engaged in the study of mathematical physics. Finally he received his degree at Breslau in 1858 and at once became a docent in the University there. After further teaching experience in a school of forestry in Neue Eberswalde, and at the Polytechnicum in Karlsruhe, he obtained in 1876 the professorship at Tübingen which he held till his death in 1898. Lothar Meyer was justly celebrated as a teacher, and his best-known book *Die Modernen Theorien der Chemie*, first published in 1864, remained the standard work upon chemical generalizations till the rise of physical chemistry in the early nineties.

Mendelejeff.—Dmitrij Ivanovitch Mendelejeff was born at Tobolsk, Siberia, in 1834. His chemical studies were begun in

Petrograd and a docentship there was followed by a year in Heidelberg between 1860 and 1861. It was altogether characteristic of the man that during this time he did not work in the university laboratory but set up a small one of his own. In 1863 he obtained a professorship in the technical school at Petrograd, which he exchanged three years later for one at the university, where he remained till 1890. In 1893 he was made director of the Bureau of Weights and Measures. He died in 1907.

			Ti = 50	Zr = 90	? = 180
			V = 51	Nb = 94	Ta = 182
			Cr = 52	Mo = 96	W = 186
			Mn = 55	Rh = 104,4	Pt = 197,4
			Fe = 56	Ru = 104,4	Ir = 198
		Ni =	Co = 59	Pd = 106,6	Os = 199
H = 1			Cu = 63,4	Ag = 108	Hg = 200
	Be = 9,4	Mg = 24	Zn = 65,2	Cd = 112	
	B = 11	Al = 27,4	? = 68	Ur = 116	Au = 197 ?
	C = 12	Si = 28	? = 70	Sn = 118	
	N = 14	P = 31	As = 75	Sb = 122	Bi = 210
	O = 16	S = 32	Se = 79,4	Te = 128 ?	
	F = 19	Cl = 35,5	Br = 80	J = 127	
Li = 7	Na = 23	K = 39	Rb = 85,4	Cs = 133	Tl = 204
		Ca = 40	Sr = 87,6	Ba = 137	Pb = 207
		? = 45	Ce = 92		
		? Er = 56	La = 94		
		? Yt = 60	Di = 95		
		? In = 75,6	Th = 118 ?		

MENDELEJEFF'S FIRST TABLE

Mendelejeff was of a highly original turn of mind and his great book, *The Principles of Chemistry*, has furnished investigators with a veritable mine of suggestive ideas for a generation, the author always having opinions of his own concerning even those points usually considered the most fixed and stereotyped. Among other things he was an irreconcilable opponent of the theory of electrolytic dissociation and hardly less heterodox in many of his other views.

There was for a time a good deal of feeling between the friends of Lothar Meyer and those of Mendelejeff upon the question of priority in the discovery of the periodic law. Such questions, however, need not detain us here. The fundamental idea, that the properties of the elements are a periodic function of their atomic weights, had been a slow growth, to which these two men independently gave the permanent form of expression. In 1882 the Royal Society conferred the Davy medal upon both in recognition of this fact, and we can well follow the spirit of their compromise. There is documentary evidence that Lothar Meyer had put in writing an arrangement of the elements as early as 1868. His first printed communication on the subject, however, was published in 1870, and contains a reference to the first paper by Mendelejeff.

	I. Gruppe	II. Gruppe	III. Gruppe	IV. Gruppe	V. Gruppe	VI. Gruppe	VII. Gruppe	VIII. Gruppe zur I. Gruppe übergehend
Typische Elemente	H = 1 Li = 7	Be = 9.4	B = 11	C = 12	N = 14	O = 16	F = 19	
Erste Periode { Reihe 1 » 2	Na = 23 K = 39	Mg = 24 Ca = 40	Al = 27.3 — = 44	Si = 28 Ti = 50?	P = 31 V = 51	S = 32 Cr = 52	Cl = 35.5 Mn = 55	Fe = 56, Co = 59 Ni = 59, Cu = 63
Zweite Periode { Reihe 3 » 4	(Cu = 63) Rb = 85	Zn = 65 Sr = 87	— = 68 (? Yt = 88?)	— = 72 Zr = 90	As = 75 Nb = 94	Se = 78 Mo = 96	Br = 80 — = 100	Ru = 104, Rh = 104 Pd = 104, Ag = 106
Dritte Periode { Reihe 5 » 6	(Ag = 106) Cs = 133	Cd = 112 Ba = 137	In = 113 — = 137	Sn = 118 Ce = 136?	Sb = 122 —	Te = 128? —	J = 127 —	— —
Vierte Periode { Reihe 7 » 8	— —	— —	— —	— —	— Ta = 182	— W = 184	— —	Os = 199?, Ir = 198? Pt = 197, Au 197
Fünfte Periode { Reihe 9 » 10	(Au = 197) —	Hg = 200 —	Tl = 204 —	Pb = 207 Th = 232	Bi = 208 —	— Ur = 240	— —	
Grenzform der O-Verbindung	R_2O	R_2O_2 oder RO	R_2O_3	R_2O_4 od. RO_2	R_2O_5	R_2O_6 od. RO_3	R_2O_7	R_2O_8 oder RO_4
Grenzform der H-Verbindung			(RH_5?)	RH_4	RH_3	RH_2	RH	

MENDELEJEFF'S SECOND TABLE

This had appeared late in 1869, while a somewhat later one printed in August, 1871, contained a long and complete exposition of his system. In it he pointed out in great detail how the position of an element in the table furnishes a guide for predicting the physical and chemical properties, not only of the element itself but also of its compounds, and that this can sometimes be done with quantitative accuracy.

The modern student is so familiar with the periodic table that it is quite superfluous to point out the fundamental details upon which Mendelejeff here laid great emphasis, such as the significance of the various series and groups, the progressive changes in electrochemical character and in valence as we pass from one group to the next, the difference between the odd and even series, the peculiarities of the eighth group, and other special features.

What impresses the reader of the present day is the thoroughness with which the author supports his view that practically everything about an element is in some way dependent upon its position in the system. It seems difficult to give an idea of this quality without a quotation from the original, and the following discussion concerning the position of indium will serve the purpose. Reference to the table will assist in following the reasoning.

"Since the atom analogs of indium, Cd and Sn, are easily reducible (even from their solutions by zinc) it must also be possible to obtain indium in this way. Since Ag (seventh series, first group) is more difficultly fusible than Cd, and the same holds true of Sb as compared with Sn, it follows from the atom analogy Ag, Cd, In, Sn, Sb that indium must be more fusible than Cd. It melts at 176°. Ag, Cd and Sn are white (of a grayish white color). These properties also belong to indium. Cd is specifically lighter than Sn, consequently indium must have a lower specific gravity than the mean between Cd and Sn. In reality this is so. Cd = 8.6, Sn = 7.2., consequently the specific gravity of In must be less than 7.9. The observed value is 7.42. Since Cd and Sn oxidize at red heat but do not rust in the air, these properties must also belong to In although in a less degree than to Cd and Sn, because Ag and Sb oxidize with still greater difficulty. Everything above mentioned agrees with the experiment. The same conclusions are reached by comparing In with Al and Tl. The specific gravity of Al is 2.67 and of Tl 11.8. The mean is 7.2.

"We may now pass to the properties of the oxides and the reactions of the salts. Indium and its atom analogs belong to the *odd* series, therefore the higher oxides cannot be strong bases. The basic character must be weaker in In_2O_3 than in CdO and Tl_2O_3 but stronger than in Al_2O_3 and in SnO_2. These conclusions are confirmed by the following facts. These oxides of Al and Sn dissolve in alkali to form definite compounds whereas the oxides of Cd and Tl are insoluble in alkali. Hence In_2O_3 dissolves in alkali without forming a definite compound. The oxides of Cd, Sn, Al, and Tl are difficultly fusible powders just like In_2O_3. The hydrate of In_2O_3, as might be expected, forms a colorless jelly. The oxides Al_2O_3 and SnO_2 are readily precipitated from their solutions by barium carbonate. So also In_2O_3. Hydrogen sulphide precipitates Cd and Sn from acid solutions, consequently indium is also precipitated. All these reactions have been confirmed by experiment."

Mendelejeff's Predictions.—Minute classification of this kind, however, was not enough for Mendelejeff and he ventured boldly

PERIODIC SYSTEM

	Group 0	Group 1	Group 2	Group 3	Group 4	Group 5	Group 6	Group 7	Group 8
Type Compound		E_2O	EO	E_2O_3	EH_4, EO_2	EH_3, E_2O_5	EH_2, EO_3	EH, E_2O_7	EO_4
Valence	0	+1	+2	+3	−4, +4	−3, +5	−2, +6	−1, +7	+8
Periods 1	He, 4	Li, 7	Be, 9.1	B, 11	C, 12	N, 14	O, 16	F, 19	
2	Ne, 20.2	Na, 23	Mg, 24.3	Al, 27.1	Si, 28.3	P, 31.0	S, 32.1	Cl, 35.5	
3	A, 39.9	K, 39.1	Ca, 40.1	Sc, 44.1	Ti, 48.1	V, 51.0	Cr, 52.0	Mn, 54.9	Fe, 55.8; Ni, 58.7; Co, 59
4	....	Cu, 63.6	Zn, 65.4	Ga, 69.9	Ge, 72.5	As, 75	Se, 79.2	Br, 79.9	
5	Kr, 82.9	Rb, 85.4	Sr, 87.6	Y, 88.7	Zr, 90.6	Cb, 93.5	Mo, 96.0	....	Ru, 101.7; Rh, 102.9; Pd, 106.7
6	....	Ag, 107.9	Cd, 112.4	In, 114.8	Sn, 118.7	Sb, 120.2	Te, 127.5	I, 126.9	
7	X, 130.2	Cs, 132.8	Ba, 137.4	La, 139.0	Ce, 140.3	Nd, 144.3	....	....	
8	....	....	....	....	....	....	....	....	
9	....	....	....	Yb, 173.5	....	Ta, 181.5	W, 184	....	Os, 190.9; Ir, 193.1; Pt, 195.2
10	....	Au, 197.2	Hg, 200.6	Tl, 204.0	Pb, 207.2	Bi, 208	....	....	
11	Nt, 222.4	....	Ra, 226.0	....	Th, 232.4	....	U, 238.2		

A MODERN TABLE OF THE PERIODIC SYSTEM

into prophecy. His table contained several vacant spaces, and from the position of these he made bold to predict not only the atomic weights of new elements which might be expected to occupy them, but also their physical properties and those of their compounds. These predictions found striking confirmation in the discovery of scandium, which showed the properties of his hypothetical "ekaboron," gallium with those of his "ekaluminum," and germanium with those of his "ekasilicon." In the last case the predictions were verified with an accuracy almost startling, as shown in the following table:

	Properties predicted for "ekasilicon"	Properties observed in germanium
Atomic weight	72.0	72.3
Specific gravity	5.5	5.469
Atomic volume	13.0	13.2
Specific gravity of oxide	4.7	4.703
Boiling point of chloride	100°	86°
Specific gravity of chloride	1.9	1.887
Boiling point of ethyl compound	160°	160°
Specific gravity of ethyl compound	0.96	1.0

Agreement of just this marvelous kind was scarcely observed in other cases and must here be ascribed in some measure to chance, but it made a striking appeal to popular attention and doubtless did much to hasten the adoption of the periodic law as one of the fundamental chemical generalizations.

Meyer's Atomic Volume Curve.—We are indebted to Lothar Meyer for a particularly happy graphic representation of the law which he first developed in his famous atomic volume curve. The principle is the familiar one of plotting, by means of rectangular coördinates, the atomic volumes of the elements against their atomic weights. The resulting curve shows numerous points of interest. It takes the form of a series of well-marked 'waves,' upon which similar elements are found occupying analogous positions. Ascending slopes contain electro-negative elements and descending slopes the electro-positive ones. The chief interest here, however, is perhaps less in the curve itself than in the suggestive method of representation, which can of course be employed equally well for plotting any

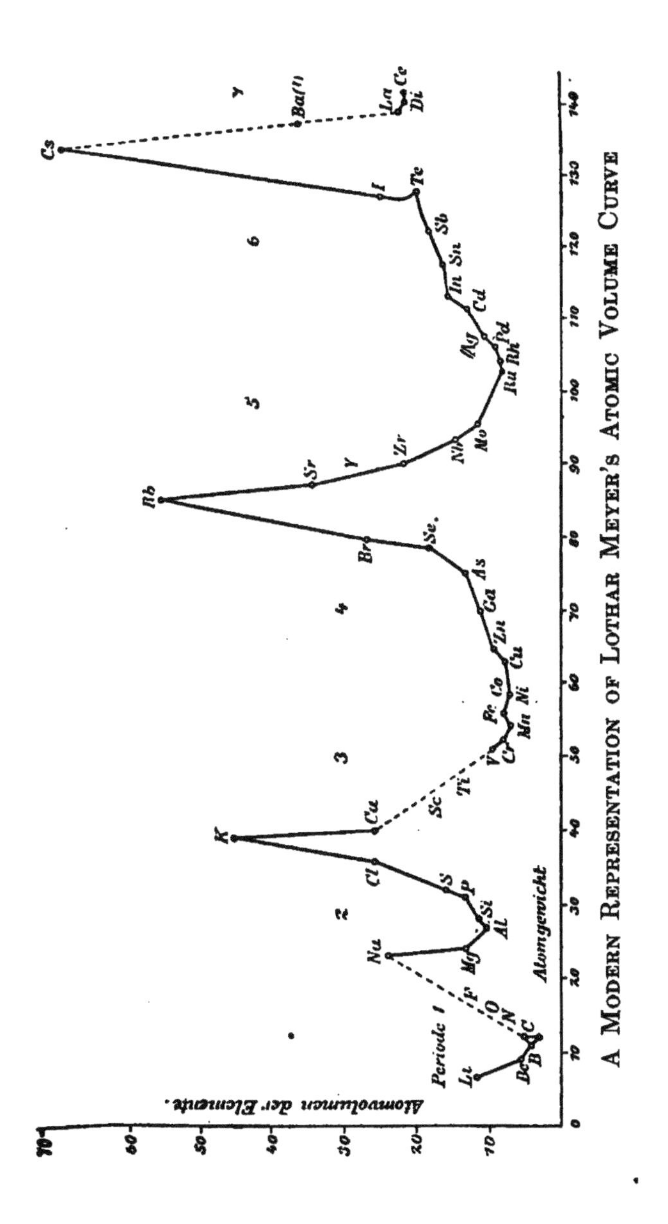

A Modern Representation of Lothar Meyer's Atomic Volume Curve

er physical property as a function of the atomic weight. en we do this, whether the property be hardness, comssibility, boiling points of analogous compounds, or what we , we usually find the same kind of recurrent periodicity in the perties concerned.

ater Developments.—Ingenious chemists have frequently sugted other representations, some involving a radical rearrangeat of the groups, others concerning themselves only with the n. Most of these have, however, made no general appeal the popular imagination, and the table of Mendelejeff is still basis of discussion whenever the periodic relations of the nents are in question. There are still serious anomalies. In er to bring similar elements into the same groups it is essary to transpose, for example, the natural position of alt and nickel, of potassium and argon, of iodine and tellun. Scores of careful investigations undertaken in the hope removing these contradictions have only served to accene them. On the other hand, the law has been splendidly licated by some particularly severe tests. When Rayleigh Ramsay discovered a whole family of new elements in the gases of the atmosphere the question at once became rife o where places for so many elements could be found in the odic classification. It soon proved, however, that these s formed a new group of their own with valence zero, whose tence could hardly have been predicted, but which, when once ized, harmonized entirely with the spirit of the law. So too, n radium was discovered great interest was aroused as to re an element of such marvelous properties would find a e, but as soon as a determination of the atomic weight became sible this element found the natural position below barium which its chemical properties entitled it.

t is natural to ask what bearing, if any, this generalization upon the question of the nature of the elements and other culations in the spirit of Prout's hypothesis. Mendelejeff iself with surprising conservatism and in spite of all the iarkable relationships he had discovered, declined to draw any clusions, and did not believe that his law necessarily threw more light upon this perplexing question than such a generaliion, for example, as that of Boyle. We shall soon see that the modern electron theory illuminates the periodic system from another point of view, but this must be postponed for consideration in its proper place.

Literature

Stas published *Recherches sur les Rapports Réciproques des Poids Atomiques*, Brussels, 1860, and *Nouvelles Recherches sur les Lois des Proportions Chimiques, sur les Poids Atomiques et leurs Rapports Mutuels*, 1865.

The papers of Lothar Meyer and Mendelejeff are reprinted in No. 68 of the *Klassiker*, while No. 66 gives the early work of Döbereiner and Pettenkofer. See also Garrett, *The Periodic Law*, London, 1909, and Venable, *Development of the Periodic Law*, Easton Publishing Company, 1896. One of the Chemical Society Memorial Lectures is devoted to Lothar Meyer, and one of Thorpe's *Essays* to Mendelejeff.

LOUIS PASTEUR

1822-1895

Reproduced from the "Chemical Society Memorial Lectures" by the kind permission of the Council of the Society.

CHAPTER XVII

BUNSEN, BERTHELOT, AND PASTEUR

Having traced the development of the science down to 1870, when the more fundamental of our modern views may be considered as definitely established, it seems appropriate to turn back and give an account of the work of three great chemists to whom little allusion has yet been made. They may all be considered as younger contemporaries of Liebig and Wöhler, but they entered far less than these men into the theoretical controversies of their time.

The history of a nation should be a record of the development of the national character, but it is most easily written as a chronology of sieges and battles. So, too, the history of a science should record the progress of the race toward knowledge in some special field, but it easily becomes an account of dominant theories as they have superseded and conflicted with each other. Here and there, however, there arise great men whose life is not spent in the service of any theory, but who rather provide science with those facts to which all theories must conform.

Bunsen, Berthelot and Pasteur exemplify this. In spite of wide differences in temperament and in their fields of activity, they resembled each other in their aversion to all unessential hypothesis, in the fundamental value of their work to humanity, and in the energy and devotion which they gave to that service.

Bunsen.—Robert Wilhelm Bunsen was born in Göttingen, May 31, 1811, his father Christian Bunsen being the librarian of the University. After attending the gymnasium at Holzminden, he entered the university of Göttingen in 1828 and received the doctorate two years later, presenting a Latin thesis upon different types of hygrometers. Bunsen spent the winter of 1832–3 in Paris, and afterward traveled quite extensively, making longer stops in Berlin and Vienna. The year 1834 found him established as a docent in the University of Göttingen. In

1836 he succeeded Wöhler at Cassel, and three years later accepted a professorship at Marburg, which he retained till 1851. In that year he made a brief change to Breslau and then in 1852 accepted the professorship at Heidelberg which he retained till his retirement from active service in 1889. He died there August 16, 1899.

Cacodyl.—As soon as Bunsen was settled at Cassel he began some investigations upon the organic compounds of arsenic which would alone have assured him recognition. Many years before, a French chemist named Cadet had distilled arsenious oxide with potassium acetate and obtained a liquid of a terrible odor which was not only intensely poisonous but also spontaneously inflammable. It is not surprising that these properties protected the substance from further investigation for many years. Bunsen, however, now attacked the problem, and found that the chief component of this dreadful liquid was an organic compound of arsenic, and that it had many of the properties of a metallic oxide. We now write the reaction which accounts for its formation as:

$$4CH_3COOK + As_2O_3 = 2CO_2 + 2K_2CO_3 + \left(\begin{matrix} CH_3 & \\ & As \\ CH_3 & \end{matrix}\right)_2 O$$

In Bunsen's time the determination of organic structure was not possible, but he recognized in the complex $C_4H_{12}As_2$ what was essentially a complex metal, to which he gave the name of cacodyl, Kd, on account of the terrific odor of most of its compounds. With acids the oxide formed salts:

$$KdO + 2HCl = KdCl_2 + H_2O$$

the chloride, bromide, cyanide, etc., and when such a salt was treated with a metal like zinc, halogen was removed:

$$KdCl_2 + Zn = ZnCl_2 + Kd$$

and what in those days passed for the free radical was liberated. Of course we now know that the resulting compound, like dicyanogen, has twice the molecular weight of the true radical, but at the time this discovery was made, it was seized upon by Berzelius as one of the most important arguments ever furnished for the truth of the radical theory—ranking in this respect with e cyanogen of Gay-Lussac and the benzoyl radical of Liebig d Wöhler.

In the successful investigation of these substances Bunsen tablished his reputation once for all as a master of chemical anipulation, but an explosion of cacodyl cyanide cost him the ght of his right eye, and weeks of illness resulted from inhaling fumes. From this time on Bunsen devoted himself exclusively work in the inorganic field.

Gas Analysis.—About 1838 he undertook the investigation blast-furnace gases, with the direct object of bringing about the ost efficient use of the fuel. He accomplished this, but the udy led to a revision and expansion of the whole subject of gas alysis, which proved classic, and Bunsen's one book *Gas-etrische Methoden* is still a work of reference in this subject. ore rapid methods have since been devised but none exceeded s in accuracy for many decades.

Geological Studies in Iceland.—In 1846 Bunsen spent three four months in Iceland where he devoted himself to the study the rocks, and took a great interest in the action of the geysers. ompetent judges have referred to his work on the Icelandic cks as laying the foundation of modern petrology, while his ethod of attacking the problem of geyser action seems of suffi-ent general interest to warrant a brief résumé here. Before is time most geologists had believed that the geyser water was lcanic. Bunsen, however, was able to prepare water like it composition by boiling rain water with the local rocks, and me to the conclusion that it was really of surface origin. e also found that only the alkaline springs dissolved silica and nly these formed geysers. A geyser according to Bunsen mply represents a deep tube or fissure in the earth in which lkaline water has settled and which is heated unequally by the otter rocks around. In such a long narrow and nearly vertical ube there will be no free circulation of the water so that this will general be considerably hotter toward the bottom than at the p. On the other hand, the pressure of the water column raises e boiling point of the water very markedly as the depth in-eases. By sinking self-registering thermometers at various epths into the tube, Bunsen found that a few minutes before the uption the temperature at several places was very close to the boiling-point at that depth. If we now assume that this boilin point is reached at any place the first effect must be to lift th water column above the point where steam is first formed. Th pressure being once reduced, the water further down now find itself heated far above its boiling-point, and bursts into explosiv ebullition, until the whole mass is discharged violently into th upper air. The water of course gradually falls back into it basin, refills the tube and the process repeats itself as befor In cooling and drying also, the water deposits some of the silic it held in solution while superheated, and this accounts for th building up of the crater, and the silicious lining of the geyse tube.

The Photochemical Investigations—The twelve or fiftee years just following 1850 probably represent the most productiv of Bunsen's life. About 1852 he introduced the processes o iodimetry into volumetric analysis, and not long after, in connec- tion with his student Roscoe, he took up the quantitative stud of the action of light upon chemical reactions. That chosen to especial study was the formation of hydrochloric acid from its elements. Hydrogen and chlorine were mixed in molecular pro- portions, and subjected in a specially designed apparatus to the action of light of known intensity for varying periods of time. It was found that the quantity of hydrochloric acid formed was proportional to the intensity of the light and the time of exposure, and, what was perhaps of more general interest, that the light absorbed in passing through such a reacting medium was propor- tional to the chemical change produced, so that the photo- chemical absorption followed the same laws as ordinary absorp- tion. Bunsen also called attention to a phenomenon which has not even yet been quite satisfactorily explained, namely the preliminary exposure which is required before the reaction acquires a constant velocity, and which must be repeated when- ever action has been arrested for a few minutes. This he called the period of "photochemical induction." This brief descrip- tion does no justice to the extent and quality of the work, which Ostwald has well characterized as the model of all that a physico- chemical investigation should be.

The Spectroscope.—In 1854 Kirchoff came to Heidelberg as professor of physics and he soon began to work with Bunsen upon

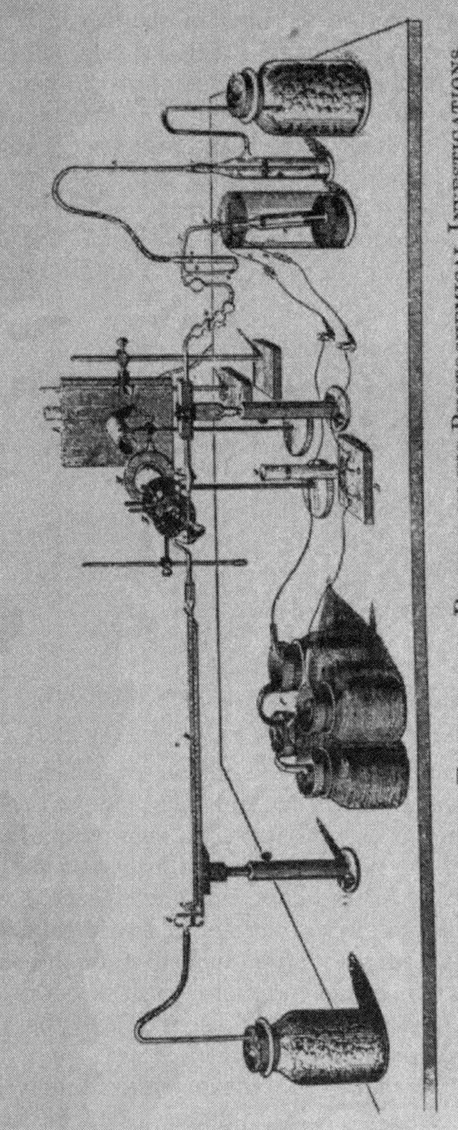

SOME OF THE APPARATUS EMPLOYED BY BUNSEN IN HIS PHOTOCHEMICAL INVESTIGATIONS

problems connected with optics. For some time previously Bunsen had been paying attention to flame tests in qualitative analysis, and had been in the habit of showing that flames in which several elements are being simultaneously volatilized can conveniently be resolved by looking at them through a prism, when each color stands out separately. The narrower the flame the sharper the definition, so that the next logical step was to allow the flame to shine through a narrow aperture, to direct the rays by a telescope in a parallel stream upon a prism, to view the spectrum thus produced by another telescope, and to enclose the prism itself in a box to protect it from diffused light. The

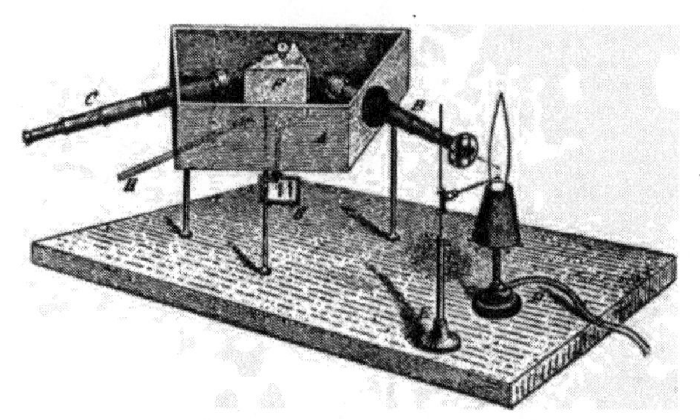

THE FIRST FORM OF THE SPECTROSCOPE

result was the spectroscope, an instrument which has added constantly to human knowledge from that day to this. In the hands of Kirchhoff it demonstrated the absorption of radiations by the vapor of the same substances which emit them, and in this way accounted for the dark Fraunhofer lines in the sun's spectrum by the presence of certain elements as gases in its atmosphere. Bunsen soon after applied it to the analysis of the water of certain springs, and the result was the discovery of the new elements rubidium and caesium, and the characterization of their important compounds.

Bunsen as a Teacher.—The above represent only a few of Bunsen's more celebrated investigations, but he is scarcely less well known for many characteristic and original tricks of manipulation and ingenious pieces of laboratory apparatus whose efficiency is only equalled by their extreme simplicity. To chemists it is almost superfluous to name them: the Bunsen burner, the battery, the ice calorimeter, the Bunsen valve (a slit in a rubber tube), the photometer (a grease spot on a piece of paper), all remind us of the ingenuity and practical sense of the master. Last but not least, Bunsen was a great teacher, only rivalled by Liebig and Wöhler in the number and distinction of the students whom he attracted to his laboratory. Here his simplicity of character, fatherly interest in his students, and unbounded sense of humor made him no less beloved than admired by all. It is recorded that in 1856 the following chemists of future eminence were at one time enrolled as students in his laboratory: Beilstein (of the great *Handbuch*), Lothar Meyer (of the periodic law) Quincke (long professor of physics in Heidelberg, and an authority on surface tension), Landolt (best known by his portentous *Tables*), Roscoe (of the Roscoe-Schorlemmer text-book), Volhard (whom students associate with two well-known analytical processes), and Adolf Baeyer (successor of Liebig in Munich and the hero of the indigo synthesis). At the same time Kekulé was a docent in the University. Here certainly must have existed a scientific atmosphere.

Berthelot.—Marcellin Pierre Eugène Berthelot was born in Paris October 29, 1827. He early attended the *Collège Henri IV*, and in 1846 won a prize in philosophy open to the competition of students in all the *lycées* of France.

He next attended the *Collège de France* where he began the study of medicine, but gradually interested himself more and more in chemistry, coming under the influence of Pelouze, Dumas, Claude Bernard, Regnault and Balard. He finally became assistant to the last named, and in this position was fortunate enough to find much time for his own researches. In 1859, Berthelot was made professor at the *École Supérieure de Pharmacie*, a position which he held till 1876, although in 1860 he also accepted a professorship at the *Collège de France* which had been created especially for him and which offered opportunities for research altogether unequalled in France at that time. A laboratory was set apart for his especial use, and the routine duties consisted only of forty public lectures a year, with no obligation to hold examinations, and perfect freedom to make the lectures whatever he chose. As a matter of fact he made them largely accounts of his own researches.

Berthelot became a member of the Academy of Medicine in 1863, of the Academy of Sciences in 1873, and was made inspector of higher education in 1876. Almost alone among scientific men he also reaped high political honors, being created a senator for life in 1881, minister of public instruction 1886–7, and minister of foreign affairs from 1895–6. He died suddenly in Paris on March 18, 1907.

Organic Syntheses.—Almost the earliest of Berthelot's publications revealed the general tendency of his subsequent work. In 1851 he passed alcohol, acetic acid and other simple substances through hot tubes and by such pyrogenetic reactions prepared benzene, phenol and naphthalene. A very indirect synthesis of acetic acid was already known, so that this reaction at once opened to synthesis whole classes of substances hitherto unattainable in this manner.

Ever since Wöhler's preparation of urea from ammonium cyanate in 1827, the preparation of isolated compounds from the elements had succeeded here and there, but these researches of Berthelot (which he soon extended further) at last began to justify the hope felt by every chemist that it may sometime prove possible to prepare all the complex compounds met with in nature by laboratory processes. Berthelot devoted himself to the realization of this ideal with zeal, and it is worth recording that he was the first to use the word synthesis in this connection. One of his next successes was the preparation of the fats in glass by the action of glycerol upon the fatty acids. This formed a part of an extensive investigation of glycerol which resulted not only in the discovery of many important derivatives like the allyl compounds, but also in the recognition that this important substance is a tri-atomic alcohol. Berthelot expressed this in the form that glycerol stands to ordinary alcohol in the same relationship that phosphoric acid stands to nitric. We have already seen (page 158) how this idea was extended when Wurtz discovered the glycols. Meantime Berthelot's studies in the terpene series served to connect oil of turpentine with camphene and camphor, and work on the sugars interested him in fermenation. Here he discovered in yeast the *invertin* which has the roperty of hydrolysing cane sugar, and he expressed the opinion hat the transformation of sugar to alcohol was doubtless due to ome other enzyme contained in the yeast. This suggestion, owever, had to wait for nearly forty years before it was justied by Buchner's discovery of *zymase.*

Berthelot next continued his synthetic studies by verifying Faraay's observation that alcohol can be obtained from ethylene hrough ethyl sulphuric acid. Having also worked out his well nown preparation of methane from carbon bisulphide, he btained from the former first methyl chloride and then methyl lcohol, thus opening to the theoretical possibilities of synthesis ll the substances which can be prepared from these two alcohols. 'his investigation occupied about ten years and the results vere published in an important book *Chimie Fondée sur la Synthèse.*

Studies on Ester Formation.—After 1860 Berthelot took less nterest in the development of pure organic chemistry, turning is attention more and more to the forces which govern chemical eactions in general. In 1862 there appeared a very famous aper published in collaboration with Péan de St. Gilles concerning the velocity of esterification. It was found that when an cid and alcohol are brought into contact, the reaction etween them never reaches completion but stops at a definite quilibrium point, and it was further found that, at any moment, he quantity of ester formed is proportional to the products of he active masses of acid and alcohol present. This is, as we see, ssentially our modern mass action law applied to a single reaction, and the statement of such a relationship in mathematical orm represents one of the first efforts of its kind.

Acetylene.—Meantime Berthelot was continuing his synthetic tudies by brilliant contributions to the chemistry of acetylene. Although not the discoverer of this substance, he gave it its present name, and established many of its singular properties, among others its preparation from the elements in the electric arc and its polymerization to benzene when strongly heated. This last observation led to the syntheses of many other complex hydrocarbons by pyrogenetic methods. Among these were styrene, naphthalene and acenaphthene.

Studies in Thermochemistry.—Even in his earliest work Berthelot had interested himself in the quantity of heat evolved or absorbed in a chemical reaction, but after 1869 he made a most extensive study of the subject with particular reference to the combustion of organic compounds in his calorimetric bomb, also including many other reactions, so that most of our data on this subject are due either to him or to Julius Thomsen of Copenhagen who also made studies of this kind his life-work. The investigation involved countless observations and an immense amount of computation. Among the results of general interest was a thorough confirmation of the important principle (not discovered by Berthelot) that in any chemical transformation the amount of heat evolved or absorbed by a given series of reactions depends only upon the initial and final states of the substances concerned and not at all upon the steps involved; thus if we start with hydrochloric acid, ammonia, and water at a given temperature, and end with a solution of ammonium chloride in water it makes no difference whether we first allow the dry gases to react, and then dissolve the product in water, or whether we dissolve the hydrochloric acid and ammonia separately in water and then mix the solutions.

Berthelot also derived from his researches a supposed law to which he attached great importance but which has not stood the test of modern criticism. This he called the *principle of maximum work.* It states that a chemical reaction always takes place with the production of those substances whose formation involves the greatest evolution of heat. Although this has been found to be thermodynamically unsound, it serves as a practical rule which holds true in a great majority of cases.

Work on Explosives.—During the Franco-Prussian war, Berthelot was a member of the scientific committee organized for the defence of Paris and became interested in explosives, dealing not only with the practical side, but also studying the nature of explosive reactions in general. The details can, of course, receive no adequate description here. One result, however, deserves mention. Berthelot studied with especial care the propagation of explosions in gas mixtures when these are confined in long tubes, and found that in such cases the velocity of transmission is independent of the pressure and of the diameter of the tube, being strictly characteristic for the gas mixture concerned. To this Berthelot gave the name of the *explosion wave.*

In 1883 a tract of land originally belonging to the palace at Meudon near Paris was placed at Berthelot's disposal, and he established there an annex to the organic laboratory of the *Collège de France* which he used as an agricultural experiment station, and during his later years spent much time there carrying on researches especially devoted to the nitrogen supplies of the growing plant.

Historical Studies.—In addition to all his chemical and political activities Berthelot still found time for original historical studies. He visited Egypt in 1869 to attend the opening of the Suez Canal, and the expedition served to stimulate his curiosity concerning the chemical knowledge possessed by the ancients, and the origin of alchemy. From this time on, as opportunity offered, he devoted much effort to the collection and translation of rare manuscripts, even those in Eastern languages, to the comparison and editing of alchemistic texts, and to the analysis of old coins and other utensils brought to light by the labors of archæologists. His books, *Les Origines de l'Alchimie* and *La Chimie au Moyen Âge*, as well as several other less-known writings, give an account of his labors in this field. Indeed Berthelot's literary productivity is no less remarkable than the great quantity of his experimental work, more than twenty-five imposing volumes having come from his pen in addition to his contributions to the scientific journals.

Mental Attitude.—Detractors have criticised his experimental work on the ground of lack of thoroughness and even of accuracy, but we have to remember that Berthelot was interested primarily in the larger features of the problems which he attacked. The isolated fact or observation he valued, not for itself, but for the light it threw upon the main question, and so he was unwilling to stop and establish the best conditions for every reaction, or sometimes even to verify his compounds by the most careful analyses. In fact, he sometimes took the liberty of genius, to rely on his intuitions, and he seldom had to regret his courage. Berthelot always brought an entirely original point of view to the solution of his problems, and this led to a disregard of the work of others which sometimes bordered on injustice. This independence expressed itself characteristically in his formulæ. For one devoting himself to organic synthesis, the modern student is apt to think of our structural formulæ as an indispensable help and guide, but Berthelot only adopted them in the last years of his life, recklessly using the old equivalents instead of the atomic weights, and thereby attaining formulæ essentially empirical, but modified in ways of his own in order to bring out special relationships which he wished to emphasize. There is no reason to dwell on these formulæ for no one else employed them, but they illustrate the peculiar combination of originality and conservatism in Berthelot's mind, and his innate aversion to the hypothetical element involved in all ideas of atoms and molecular structure. When a friend once told Berthelot that he need not take the atoms so seriously, that using them as aids to thought need imply no belief in their objective existence, he replied with a trace of bitterness, "Wurtz has seen them!"

This aversion to hypothesis was a part of the very philosophy which in Berthelot's case seems to have been the compelling inspiration of his work. In student days he formed an intimate and lifelong friendship with Ernest Renan and both became thoroughly imbued with the spirit of religious scepticism so common in France at that time, and expressed by the latter in his *Life of Jesus*. Berthelot, on his side, seems to have assumed the task of showing that all the remarkable transformations of the organic world are due to the play of simple chemical and mechanical forces acting in a mechanical way. As he himself expressed it, "It is the object of these researches to do away with *life* as an explanation, wherever organic chemistry is concerned." Strange as such an aim now appears to us, and little as it would now seem to prove even were the whole contention conceded, we find in Berthelot an interesting and unusual case of a life producing a wealth of positive, constructive results when inspired by a spirit of negation.

Pasteur.—Louis Pasteur was born in Dôle, France, December 27, 1822. Not long after, the family removed to Arbois and there most of Pasteur's youth was spent. Having attended the colleges at Arbois and at Besançon, he became *bachelier ès lettres* in 1840 and *bachelier ès sciences* in 1842, being set down as *mediocre* in chemistry—another interesting commentary on academic standards. In 1843 he entered the *École Normale* at Paris where, three years later, he became assistant to Balard with opportunities for independent investigation. His first appointment was to the faculty of Dijon, but he soon exchanged this position for an assistantship in Strasburg. In 1854 he was made professor at Lille and dean of the Faculty of Sciences, a position which he surrendered three years later to return to the *École Normale* in Paris, and with this institution he remained connected throughout practically the whole of his active life. Among other honors, he was made a member of the Academy of Sciences in 1866, of the Academy of Medicine in 1873, and of the French Academy in 1882. He died in Paris, September 28, 1895.

Work on Tartaric Acid.—Practically the first scientific problem which engaged the attention of Pasteur proved to be of far-reaching significance. In Balard's laboratory he had come in contact with Laurent, who interested Pasteur in the microscopic study of crystals. These studies led him to follow the work of Mitscherlich on arsenates and phosphates and to include in his observations the salts of tartaric and of racemic acids. According to Mitscherlich the double sodium ammonium salts of these acids crystallized in exactly similar forms and yet, in solution, the tartrate rotated the plane of polarized light to the right while the racemate was optically inactive. A more thorough examination by Pasteur, however, showed that the crystalline forms were not exactly alike; that the tartrate showed certain hemihedral faces which occurred only on the right side of the crystal, whereas in the case of the racemate such faces appeared upon different crystals sometimes on one side and sometimes upon the other. It occurred to Pasteur to separate the right-handed from the left-handed crystals and to examine them separately with the polariscope. He now found that the right-handed ones were nothing else than the familiar tartrate, while the left-handed crystals represented the salt of a hitherto unknown acid, exactly like tartaric in all other respects, but rotating the plane of polarized light as much to the left as tartaric acid did to the right. Now the connection between optical rotation and hemihedral forms was not new. Hauey had observed the occurrence of hemihedral faces among quartz crystals and Biot had noticed that some specimens of this mineral rotate to the right while others rotate to the left.

Herschel combined these observations showing that the geometrical form goes hand in hand with the direction of rotation. Pasteur's present observation was a distinct step in advance, for the tartrates rotate the plane of polarized light *in solution*, and it was soon to be discovered that optically active volatile organic compounds show the property even in the *gaseous* state. From this Pasteur drew the bold and correct conclusion that the molecule of such a compound is itself unsymmetrical.

He has left a pleasing account of the interest which Biot, then an old man, took in these researches:

"The announcement of the above facts naturally placed me in communication with Biot, who was not without doubts concerning their accuracy. Being charged with giving an account of them to the Academy, he made me come to him and repeat before his eyes the decisive experiment. He handed over to me some paratartaric[1] acid which he had himself previously studied with particular care, and which he had found to be perfectly indifferent to polarized light. I prepared the double salt in his presence, with soda and ammonia which he had likewise desired to provide. The liquid was set aside for slow evaporation in one of his rooms. When it had furnished about 30 to 40 grams of crystals, he asked me to call at the *Collège de France* in order to collect them and isolate before him, by recognition of their crystallographic character, the right and left crystals, requesting me to state once more whether I really affirmed that the crystals which I should place at his right would deviate to the right, and the others to the left. This done, he told me that he would undertake the rest. He prepared the solution with carefully measured quantities, and when ready to examine them in the polarizing apparatus, he once more invited me to come into his room. He first placed in the apparatus the more interesting solution, that which ought to deviate to the left. Without even making a measurement, he saw by the appearance of the tints of the two images, ordinary and extraordinary, in the analyser, that there was a strong deviation to the left. Then, very visibly affected, the illustrious old man took me by the arm and said, "My dear child, I have loved science so much all my life that this makes my heart throb."

The incident marked the beginning of a friendship which ended only with the death of Biot and proved of great advantage to Pasteur.

[1] An earlier name for racemic acid.

The idea of molecular structure in our modern sense was not developed at this time, so Pasteur was not in a position to refer back the asymmetry of the molecule to the particular atoms which are responsible as we do today. He did, however, account for the four forms of tartaric acid and worked out the three methods for splitting racemes which are still our main reliance in work of this kind. The method of mechanical selection we have already described, but Pasteur added two others. He saw that the combination of a right-handed acid, for example, with a left-handed base could not have the same properties as the compound of a left-handed acid with the same base, and acting on this idea he devised our present methods of adding to a raceme another active complex, usually involving salt formation. Pasteur also made the important observation that when *penicillium glaucum* grew in a racemate solution the right-handed form gradually disappeared, while the other was unattacked. This not only furnished a method frequently applicable for obtaining one component of a raceme, but it also demonstrated the important principle that optical opposites, in spite of their striking similarity in other respects, show marked differences as soon as physiological influences come into play.

Studies in Fermentation.—Work with microörganisms of this kind was destined later on to absorb all of Pasteur's activities. When professor at Lille he became interested in the troubles met with by a local distillery in fermenting beet sugar. The difficulty consisted in an undesired fermentation which was producing lactic acid. Study soon showed that during the process certain microörganisms appeared in the fermenting liquid which were foreign to a healthy alcohol fermentation, and that these, when placed in a fresh sugar solution, induced a renewed formation of lactic acid. The obvious conclusion was that the fermentation was *caused* by the organism. Familiar as such an idea seems to us, it was in direct contradiction to the general opinions of the time.

Although fermentative and putrefactive processes had long been familiar and one of them, alcoholic fermentation, had been practised industrially for centuries, next to nothing was really understood about the nature of the process, and this in spite of the fact that a good deal of evidence was available which might have been used for the solution of the problem. In the seventeenth century the microscope had become so far perfected that it was possible to observe the forms of yeasts and bacteria, but it was only in 1803 that L. J. Thénard ventured to declare that these organisms were the cause of the chemical action involved. While this made some impression at the time, the idea again lost ground, largely on account of the opposition of Berzelius and Liebig, whose tendency to dogmatize here had very unfortunate results. Neither Berzelius nor Liebig paid much attention to the fact that yeast is a living organism, though neither could be said to be ignorant of the fact. To Berzelius the action of yeast upon a sugar solution was a splendid example of *catalysis*, by which he understood rather more than we now ascribe to the word, regarding it as a kind of contact force. Liebig's theory was more elaborate. He reasoned that a substance is only stable when the amplitude of vibration of its atoms does not exceed a certain amount, for if these get beyond the range under which chemical affinity acts, the compound must obviously decompose. If now there be brought into contact with a reasonably stable substance another which is already undergoing putrefaction (so Liebig regarded yeast), then the escaping decomposition products of the latter act upon the atoms of the more stable one, as one tuning fork affects another of the same pitch, causing decomposition of the compound.

The view of Berzelius has been in a measure justified by recent investigation, for we now know that the yeast contains an enzyme, *zymase*, whose presence, rather than the vital processes of the yeast itself, brings about alcoholic fermentation. The theory of Liebig, however, was in reality only a crude attempt to picture the mechanism of the catalysis and it rested upon no adequate experimental evidence. It was, however, defended by its author with a vigor and obstinacy characteristic of the man, and inversely proportional to the strength of the argument—a fact which will hardly surprise any student of human nature.

Pasteur's experiments were conclusive. He showed that when fermentation occurs certain microörganisms abound in the liquid, that when these are introduced into an unfermented liquid of the same kind fermentation is at once induced, that the kind of fermentation depends upon the kind of microörganism, and finally, that when organisms are rigorously excluded no fermentation occurs.

Spontaneous Generation.—The support of this last proposition involved Pasteur in a long and bitter controversy upon the old subject of spontaneous generation which in one form or another had vexed the world for centuries. The observation that decaying matter usually abounds with all sorts of life is as old as the race, and in early days the assumption was naturally made that the animals were the product of decay, even as acute an observer as Van Helmont asserting that mice could be produced by mixing meal with dirty rags. We can forgive this to Van Helmont, but it seems well-nigh incredible that views not much less crude in principle could persist beyond the middle of the nineteenth century. It is true, of course, that modern believers in spontaneous generation confined their arguments to animals much smaller than mice, but philosophical controversies on the subject recurred with regularity, and were always especially bitter, because one side or the other invariably tried to make the topic a factor in religious discussion. Its irrelevancy here was abundantly shown by the fact that spontaneous generation figured alternately on both sides of the question. This was clearly seen by Voltaire. Those who recall the inscription above the door of the chapel at Fernay, "*Deo erexit Voltaire*," will remember that its author was sometimes willing to break a lance on the side of orthodoxy, and concerning the skeptics of his day who used the argument of spontaneous generation to justify their atheism he remarked pithily, "It is strange that men should deny a Creator and yet arrogate to themselves the power of creating eels."

Attitude toward Religion.—Unlike Berthelot, Pasteur's temperament was deeply though unaggressively religious and like Faraday (who was also a devout Christian) he never permitted doctrinal bias to influence in the slightest degree the inflexible accuracy of his experiments, realizing that natural science has certain limitations which it can never pretend successfully to pass. Its sphere is the observation and correlation of sense phenomena, whereas religious truths are not sense phenomena, and must be "spiritually discerned."

CHAPTER XVIII

ORGANIC CHEMISTRY SINCE 1860

The long controversies which ended about 1860 in the triumph of Avogadro's hypothesis and the vindication of the atomic theory had been fought out in the organic field, and had culminated in the establishment of the valence theory as the guiding principle in that branch of the science. This gave, perhaps, to organic chemistry a somewhat exaggerated importance—at any rate, the idea that chemical compounds could be visualized as groups of real atoms united by real bonds exerted a remarkable fascination, and young chemists in great numbers began to devote themselves to synthetic studies, attempting on the one hand to prepare from the elements the most complex products of nature, and on the other to make the greatest variety of new combinations in order to find the utmost limits of chemical affinity and molecular stability. The rise of the coal-tar industry and the possibility of preparing from this source so many compounds of practical utility was partly cause and partly effect of this great movement which is going on uninterruptedly at the present day.

If, however, we ask what direct contribution to the science as a whole has been made by organic chemistry since 1860 we can hardly give it so high a place. We must rather confess that this branch of the science has lived largely for itself and while it has, during that time, developed a real history of its own which is of fascinating interest to the specialist, its great historical service to chemistry culminated in the work of Williamson, Gerhardt and Kekulé.

This special history of modern organic chemistry is far too important to pass over entirely in silence, but only those influences will be considered which yielded some new fundamental idea, or disclosed the constitution of whole classes of compounds of unusual interest. The first of these great advances was made

EMIL FISCHER
1852–

(Facing page 212)

Johann Friedrich Wilhelm Adolf Baeyer
1835–1917

through the theory of the aromatic compounds advanced by Kekulé in 1865.

Kekulé's Benzene Theory.—These substances had originally received this name on account of a peculiar odor possessed by certain representatives. Later it was found that most compounds so classified exhibited certain chemical properties in common, such as ease of nitration and sulphonation, and stability toward oxidizing agents. They also contained, as a rule, relatively more carbon than substances like alcohol or acetic acid. Their structure therefore caused particular difficulties, because the high percentage of carbon in comparison to hydrogen could hardly be accounted for save by a massing of multiple bonds entirely out of keeping with the saturated behavior of the substances concerned. Gradually it became clear that most of these compounds were closely related to benzene, and the constitution of this substance thus became of fundamental importance. At last it occurred to Kekulé that a consistent explanation was to be found in the assumption that the six carbons of benzene were arranged in a ring united by alternate single and double bonds, and with a hydrogen attached to each carbon. The constant study of the aromatic compounds in all the laboratories of the world during more than fifty years has only served to confirm this hypothesis, which may now be considered one of the most thoroughly tested generalizations of science. There is, therefore, a distinct interest in Kekulé's own account of how the idea first came to his mind:

"I was busy writing on my text-book but could make no progress—my mind was on other things. I turned my chair to the fire and sank into a doze. Again the atoms were before my eyes. Little groups kept modestly in the background. My mind's eye, trained by the observation of similar forms, could now distinguish more complex structures of various kinds. Long chains here and there more firmly joined; all winding and turning with a snake-like motion. Suddenly one of the serpents caught its own tail and the ring thus formed whirled exasperatingly before my eyes. I woke as by lightning, and spent the rest of the night working out the logical consequences of the hypothesis. If we learn to dream we shall perhaps discover truth. But let us beware of publishing our dreams until they have been tested by the waking consciousness."

double bond free rotation of the atoms about a common axis is thereby rendered impossible, and when the other two bonds of each carbon atom are attached to dissimilar groups two isomers are possible, according to whether certain groups are on the same side (*cis*) or opposite sides (*trans*) of the double bond.

$$\begin{array}{l} \mathrm{H{-}C{-}COOH} \\ \quad \| \\ \mathrm{H{-}C{-}COOH} \end{array} \qquad\qquad \begin{array}{r} \mathrm{H{-}C{-}COOH} \\ \| \quad\quad\quad \\ \mathrm{HOOC{-}C{-}H} \end{array}$$

Maleic acid (cis) Fumaric acid (trans)

Such molecules are not unsymmetrical and hence cannot be optically active, they differ also in the relative positions of their substituting groups, and hence it follows that they must differ in properties and stability. This proved an admirable explanation for such isomerism as is observed, for example, in the cases of fumaric and maleic acids, of the two crotonic acids, of angelic and tiglic acids, and many more now familiar to every student of organic chemistry.

Extension of the Theory.—Van't Hoff's book was hardly published when it was attacked with the utmost violence by Kolbe, whose aversion to mechanical conceptions was well-known, and who did not hesitate to characterize the idea that the spacial arrangement of atoms in molecules could be determined as something "not far removed from belief in witchcraft and spirit rapping." From the start, however, the new idea enjoyed the powerful support of Wislecenus whose work did much to extend and verify its conclusions, as indeed has all organic work since that time; most conspicuously perhaps that of Emil Fischer on the constitution of the sugars. The fundamental idea has also been extended with time to include other elements beside carbon. In 1890 it was shown by Le Bel that when the five valencies of nitrogen are satisfied by five dissimilar groups, optical isomerism can be realized, and the same holds true for quadrivalent tin and quadrivalent sulphur. Furthermore in the cases of trivalent nitrogen Hantzsch and Werner have made out a very strong case for something akin to *cis trans* isomerism in the case of the oximes and for the diazotates.

$$\begin{array}{r} \mathrm{C_6H_5{-}C{-}H} \\ \| \quad\; \\ \mathrm{N{-}OH} \end{array} \quad \begin{array}{l} \mathrm{C_6H_5{-}C{-}H} \\ \qquad\;\; \| \\ \mathrm{HO{-}N} \end{array} \quad \begin{array}{l} \mathrm{C_6H_5{-}N} \\ \qquad\;\; \| \\ \mathrm{KO{-}N} \end{array} \quad \begin{array}{r} \mathrm{C_6H_5{-}N} \\ \| \quad \\ \mathrm{N{-}OK} \end{array}$$

Benz syn aldoxine Benz *anti* aldoxine *Syn* diazotate *Anti* diazotate

Not long after the appearance of Van't Hoff's pamphlet similar views were published by Le Bel. The two had been fellow-students in the laboratory of Wurtz at Paris, where both had thought out essentially the same idea, each without mentioning it to the other. It is pleasant to record that the question of priority in the matter cast no cloud upon their friendship.

Bivalent Carbon.—Historically the most important step in the transition from the type theory of Gerhardt to the structure theory had been the introduction of the methane type by Kekulé. This made the quadrivalence of carbon a corner-stone of the new philosophy. Nevertheless once the theory was well established, evidence began to accumulate that this quadrivalence is by no means universal. There had, of course, always been the glaring case of carbon monoxide which everyone was willing to overlook so long as it stood alone, but about 1892 the work of John Ulric Nef of Chicago and others, began to show with increasing cogency that there were several types of organic substances, notably the isonitriles RNC: and the fulminates MeONC:, where two of the bonds of carbon are apparently unemployed. The evidence is cumulative in character and therefore unsuitable for presentation here, but chemists are now for the most part well convinced that in such compounds carbon is actually bivalent.

Trivalent Carbon.—A more startling exception was first observed by Moses Gomberg of the University of Michigan in 1900. He had set out to prepare hexaphenyl ethane $(C_6H_5)_3C.C(C_6H_5)_3$, by the action of zinc on triphenyl chloromethane $(C_6H_5)_3C.Cl$, when to his surprise instead of the inert compound which analogy led him to expect, he obtained a highly reactive substance of the same empirical composition which formed addition products with a great variety of substances, even absorbing oxygen from the air to form a stable peroxide. On account of these striking properties, and in spite of a molecular weight determination, Gomberg did not hesitate to ascribe to his substance half the formula of hexaphenyl ethane $(C_6H_5)_3C$. and to give it the name *triphenylmethyl*. This conclusion seemed so unorthodox that universal interest was at once aroused, and many chemists came forward with attempts to show how the unusual phenomena observed might be accounted for in a more conventional way.

As time has gone on, however, and especially since the remarkable work of Schlenk in 1910 upon the corresponding compounds of biphenyl, it has become increasingly apparent that, in solution at least, a carbon atom attached to three benzene rings has lost practically all affinity for another carbon similarly connected, and at the same time has acquired properties of marked unsaturation toward other substances. A whole class of such compounds are now known, and they furnish convincing proof that we here have to do with substances in which carbon is truly trivalent.

Tautomerism.—Another troublesome difficulty has lain in the fact that certain compounds apparently have almost equal claims to two structural formulæ, some of their behavior being easier to explain on the one and some on the other hypothesis. The phenomenon is called tautomerism, and aceto-acetic ester is the classic example for it reacts sometimes as if it had the formula $CH_3.CO.CH_2.COOC_2H_5$ and sometimes as if its structure were $CH_3.C(OH):CH.COOC_2H_5$. It was doubtless the influence of physico-chemical considerations which led to the true solution of this difficulty, by assuming that in all such cases two substances actually are present (though perhaps in very unequal quantities) and in dynamic equilibrium with each other, that is, that they are mutually convertible with a high velocity, so that when one component is used up by a reaction the other is immediately transformed to take its place. The idea that acetoacetic ester was an equilibrium mixture had been made probable by the work of Brühl and others upon its optical properties, but a striking confirmation of this theory was furnished when in 1911 Knorr succeeded in isolating the two forms and measuring the velocity of their mutual transformation. Tautomerism then is really isomerism in which both substances are so rapidly convertible into each other that under ordinary circumstances it is impracticable to isolate either.

Special Researches.—Still more important, however, than any of these general discussions have been the great series of researches clearing up the constitution, relationships and syntheses of whole classes of compounds. The limitations of a work of this kind would hardly permit even an enumeration of them, to say nothing of any adequate appreciation. Among the greatest rank those of Emil Fischer upon the sugars carried on since 1883, upon the derivatives of uric acid from 1892, and upon the proteins since 1899, all especially important on account of the magnificent experimental work involved and secondarily on account of the physiological significance of the substances concerned. With these may be mentioned the work of Baeyer upon the constitution of benzene, and upon indigo, the latter being especially rich in the great number of subordinate problems in all branches of organic chemistry which it raised and for which it suggested solution. There must also be mentioned Victor Meyer's work upon the derivatives of thiophene beginning in 1883, and that of Wallach upon the terpenes which has progressed uninterruptedly since 1884. Last though by no means least comes the recent work of Willstätter upon chlorophyll which began in 1906.

The Coal-tar Industry.—An account of the development of organic chemistry would hardly be complete without some mention of the more important events in the history of the coal-tar industry. Up to the middle of the nineteenth century this tar had been an extremely unwelcome and troublesome by-product of the gas works. Use was found for some of it as fuel, some was used for the preservation of timber, and the lower boiling portions were employed more or less as solvents, but these uses afforded no complete or profitable employment of the material. Aniline was found in tar by Hofmann in 1843, and the discovery led this chemist to his extensive researches on amines. In 1845 his discovery of benzene in the tar made possible the preparation of aniline and similar bases in large quantities, and in 1856 William Perkin, a student of Hofmann, while studying the action of oxidizing agents upon crude aniline oil prepared a dye which he called *mauve.* Against the advice of his teacher (who thought he ought to devote his talents to pure science) Perkin withdrew from the Royal College of Chemistry and began the manufacture of this and other products upon a commercial scale. His example was soon followed by others. Fuchsin came upon the market in 1859, and between 1858 and 1866 the work of Peter Griess, another associate of Hofmann, upon the diazo compounds made possible the almost infinite combinations now known as the azo dyes. In 1867 Graebe and Liebermann showed by a fortunate reaction that alizarin, the coloring principle of the madder plant, was really a derivative of anthracene and that it could be prepared economically from this source. This discovery had a twofold influence, for an unexpected use was now found for tar anthracene, and the large acreage in France which had hitherto been cultivated for madder was made available for the production of foodstuffs. Finally the work of Baeyer upon indigo bore such fruit in the hands of Heumann and the chemists of the Badische Company that by 1894 th's important staple could be produced from naphthalene at a price permitting competition with the natural product, and somewhat later it was manufactured from this source in a quantity sufficient to meet the requirements of the world. Although dyes have proved the most important products of coal-tar industrially, countless other compounds suitable for use as remedies, perfumes, explosives, and so forth have been p pared from the same source. To the chemist the great interest of this industry lies in the fact that every step in its progress has resulted from the application of the highest class of scientific work to the problems concerned. It has proved a veritabl triumph of mind over matter.

Literature

There is an account of Kekulé's life and work in the *Chemical Society Memorial Lectures*. His principal paper on aromatic compounds is in the *Klassiker* No. 145, but the best idea of his contribution to organic chemistry is still to be derived from the first two volumes of his *Lehrbuch der Organischen Chemie*. The third volume appeared much later and was essentially perfunctory.

Ernst Cohen's *Jacobus Henricus Van't Hoff, sein Leben und Wirken*, Leipzig, 1912, is decidedly interesting. Van't Hoff's *La Chimie dans l'Espace* has appeared in several editions and is still authoritative.

Readers interested in the recent development of organic chemistry are referred to J. B. Cohen's *Organic Chemistry for Advanced Students*, 2 vols., London, 1907 and 1913.

See also F. Henrich: *Theorien der Organischen Chemie*, Braunschweig, 1912.

Both books contain extensive references to the journal literature.

William Perkin
1838–1907

(Facing page 220)

HENRI MOISSAN
1852–1907

CHAPTER XIX

INORGANIC CHEMISTRY SINCE 1860

Between 1870 and 1890 the rapid development of organic chemistry gave it such a relative prominence that the other branches of the science rather suffered in consequence. Inorganic chemistry particularly seemed to be drifting toward the discouraging position of a completed science, and some predicted for it little further growth. The points of view which fascinated organic chemists seemed lacking in the inorganic field. No other element combines with itself as carbon does, so that a structure theory seemed here impossible, even if it were not excluded by variable valence. Furthermore, the possibility of preparing new compounds limited itself almost exclusively to salts, and these lacked interest. The true key to progress would have been in the study of electrolysis, but most workers neglected this because, since the downfall of the dualistic theory of Berzelius, all connection between chemical affinity and electricity was widely regarded as illusory. For the inorganic chemist, therefore, in those days interest centered upon the refinement of analytical procedure, the discovery of new elements, and the revision of atomic weights. Such researches were of course of great value but, from the historical point of view, they leave little to record, because they introduce little which is new in the way of important principles.

The Isolation of Fluorine.—As typical of the best inorganic work of this period may be mentioned that of Moissan on the isolation of fluorine and that of the same chemist upon the electric furnace. That some unknown element was present in the fluorides was recognized by Lavoisier who put the radical of this acid in his list of elements. By this term of course he meant not our element fluorine but something corresponding to the radical of muriatic acid, in harmony with his unfortunate theory of the nature of such substances. When Davy established the elementary character of chlorine, fluorine began to appear in the list of elements, but its isolation was delayed for many years. It is so reactive that when by any operation such as electrolysis it is for a moment set free, it attacks at once the walls of the containing vessel, the electrodes, or the solvent, and so could not be isolated. Moissan, however, by using low temperatures and platinum electrodes was able in 1887 to prepare a little of the highly reactive substance, and in order to see it adopted the ingenious device of preparing vessels from transparent fluorspar, which of course is not attacked by the element. He carried on his experiments with such skill that he was able to make accurate determinations of most of the physical properties of this extraordinary substance.

The Electric Furnace.—Moissan also gave to chemistry an important piece of apparatus in his electric furnace. This is very simple in principle. Into a box of material extremely refractory to heat, usually lime, there are introduced two electrodes usually of carbon, and between these are allowed to pass electric currents of great strength. The details differ according to whether the simple effect of heat or some reducing action is desired. With the hitherto unattained temperatures made possible by this furnace Moissan was able on the one hand to reduce from their oxides many metals hardly obtainable in any other way, and to prepare in quantity a large number of interesting carbides and other substances. Among these calcium carbide, which is now prepared in this way on the large scale, has attained great practical importance, partly for the preparation of acetylene and partly for that of cyanamide, one of the more important of the newer fertilizers. Its formation also furnishes a means of obtaining nitrogen from the atmosphere, an important detail in modern national economy.

The Dissociation of Iodine Vapor.—The acceptance of Avogadro's hypothesis stimulated interest in vapor densities, especially among organic chemists, for whom a knowledge of molecular weight is of vital importance in determining structure. The method of Dumas, although accurate, required large quantities of substance, quite out of the question in most organic researches, and was also unsuitable for high temperatures. Both Hofmann and Victor Meyer therefore perfected more suitable methods and

VICTOR MEYER
1848–1897

(Facing page 222)

the latter particularly interested himself in vapor densities at high temperature, not only in the cases of compounds but also in that of elements. In 1880 when experimenting with iodine he made the important observation that above 800° the vapor expands at a rate exceeding that required by the law of Gay-Lussac, though the density did not sink to quite half the theoretical value at 1468°, the highest temperature observed. Victor Meyer's interpretation was that at low temperatures the vapor of iodine consists of molecules containing two atoms (like those of oxygen and nitrogen) while at the highest temperatures there is but one atom in the molecule. This experiment thus furnished another excellent even if late confirmation of Avogadro's hypothesis. Some other elements show similar dissociation but in no other could this be pushed so near completion.

Werner's Work on the Metal Ammonias.—In 1892 A. Werner of Zurich began the publication of a series of papers in which he developed an entirely original conception concerning the composition of the so-called metal-ammonias, which was destined to have a marked, if at present somewhat indefinite, influence upon our general conceptions of chemical combination. It is a familiar fact that many salts of heavy metals such as cobalt, nickel, copper, chromium, iron, and members of the platinum group form addition products with ammonia in a variety of proportions. Many of these compounds exhibit striking properties, especially in the matter of color. Solutions of copper salts in an excess of ammonia contain such complexes. Long before Werner began his work a great number of these compounds had been prepared and analyzed, but no one had tried to consider them seriously from a single point of view, and most of them were formulated in the helpless way in which we still write salts containing water of crystallization.

Werner proceeded to tabulate the known compounds and to prepare others so as to make his series complete, and he found that the salts of a trivalent metal, MeX_3 for example, usually combined with ammonia in all proportions from six to three, so that we have in this case the series:

$MeX_3,6NH_3$; $MeX_3,5NH_3$; $MeX_3,4NH_3$; and $MeX_3,3NH_3$.

If now we examine the chemical properties of these compounds certain remarkable relationships appear. In the first member the whole of the acid radical is ionized. This can be shown by the electrical conductivity, or by the fact that if X represents a halogen the whole of it may be precipitated by silver nitrate. In the second member, however, this is no longer the case. Here (in $MeX_3,5NH_3$) only two-thirds of the acid is ionized, in the third member only one-third, and the fourth compound is a neutral substance which does not conduct the electric current. From these facts Werner drew the conclusion that in the first compound the three acid radicals were anions while the metal and the six ammonias formed a complex cation. In the other members of the series the acid radicals took their place successively with the ammonias in the metallic complex, finally forming the neutral compound, $MeX_3,3NH_3$. The series should then be formulated according to Werner as follows:

$$[Me,6NH_3]X_3;\ [MeX,5NH_3]X_2;\ [MeX_2,4NH_3]X;\ [MeX_3,3NH_3].$$

It was possible, however, to go further. Compounds with still less ammonia could be prepared if at the same time alkali salt were added, but now the metallic complex became the anion and the series could be completed as follows:

$$K[MeX_4,2NH_3];\ K_2[MeX_5,NH_3];\ K_3MeX_6;$$

the last term representing a type of compound of which the well-known double salt $K_3Co(NO_2)_6$ is a familiar example.

It will perhaps make these ideas more concrete if we tabulate here the compounds of the series beginning with $PtCl_4,6NH_3$ and their relative conductivities:

$[Pt,6NH_3]Cl_4$	522.9
$[PtCl,5NH_3]Cl_3$	unknown
$[PtCl_2,4NH_3]Cl_2$	228.0
$[PtCl_3,3NH_3]Cl$	96.75
$[PtCl_4,2NH_3]$	0.0
$K[PtCl_5,NH_3]$	108.5
K_2PtCl_6	256.0

Here, just as in the previous example, this method of formulating the metal-ammonias brings their composition into harmony with well-known series of double salts, and makes the composition of the latter more intelligible. In most of these compounds, also, ammonia can be substituted, molecule for molecule, by water, and as the end-members of such series we get familiar salts with water of crystallization, so that the theory throws some light even on that troublesome topic.

The Coördination Number.—In comparing the cobalt and platinum series just mentioned, the reader will have noticed that regardless of the valence of the metal, and equally regardless of the nature of the substituting groups, the so-called "inner sphere" (complex ion) consists of the metal and *six* other constituents. This number cannot be the valence of the metal but it does determine how many groups are spacially combined with it. Werner calls it the *coördination number*, and while in most cases this is six, in certain other well-known series it is four, and might so far as we know have any other value.

It will be seen that this new system is admirably adapted for use as a principle of classification, and indeed it rapidly became the basis of nomenclature and guide in research in the special field of the metal-ammonias where it originated. Werner, however, has been anxious from the first that his theory should represent something more than a series of types like Gerhardt's in which the most heterogeneous compounds might be classified on the basis of their empirical composition. Now just as the value of the structure theory in organic chemistry, as a picture of real conditions, lies in the fact that it explains isomerism, so in this new theory, if the grouping of ammonias and other radicals about a central metal atom represents a real spacial arrangement, then both structure isomerism and space isomerism should be observed, and this is the fact.

Stereoisomerism in Inorganic Chemistry.—In the first place it should make a difference whether a given atom is in the inner or outer sphere, that is, such a compound as $[CoCl,5NH_3]Br_2$ should differ from $[CoBr,5NH_3]ClBr$ and this is the case. But stereoisomerism is also possible. If six radicals are grouped about a central atom two arrangements are possible, the hexagon or the octahedron. Werner decided in favor of the latter because two di-substitution products are observed, instead of three as in the case of benzene. These he formulates for example as

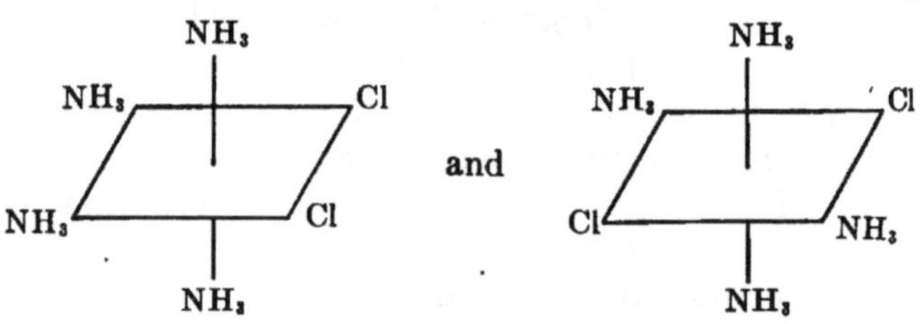

which we see is a true case of *cis trans* isomerism.

The realization of optical isomerism in this class of compounds was long delayed, partly because the preparation of compounds containing six different substituents offers experimental difficulties, but in 1911 Werner came to realize that it was not necessary to wait for the preparation of such compounds. He found that three molecules of ethylene diamine $\begin{array}{l} CH_2 - NH_2 \\ | \\ CH_2 - NH_2 \end{array}$ could replace six ammonias in complex salts, each molecule connecting, as it were, two adjacent apexes of the octahedron. Now the study of an octahedral model reveals the fact that such

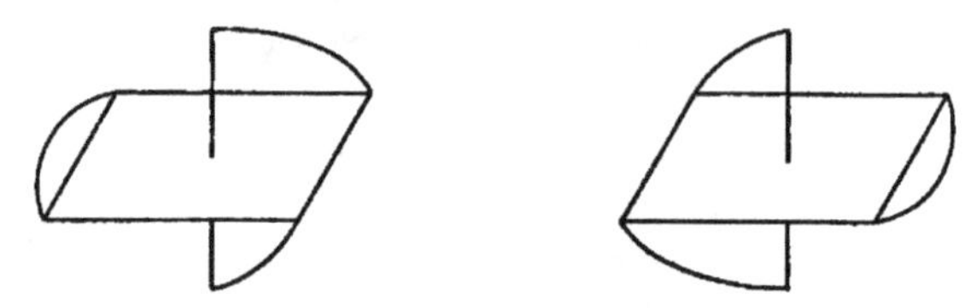

ions could occur in forms which are unsymmetrical mirror images of each other, and experiment has since shown that salts like $[Co,en_3]Cl_3$ (to use Werner's abbreviation), can be split into strongly rotating optically active components. Ethylene diamine is itself inactive, but in order to meet the possible objection that the activity of the complex might be due to the carbons of that substance, Werner, in 1914, found an inorganic radical which could be substituted for the ethylene diamine, and was able to show that the resulting compound $\left[Co\left(\begin{matrix}OH \\ OH\end{matrix}Co(NH_3)_4\right)_3\right]X_6$ was optically active though destitute of carbon.

Werner's views raise general questions of much interest. Vhat, for example, is the significance of the coördination number, nd what relation does it bear to the valence of the metal? Verner assumes that valence is more distributed than the re-resentation by individual bonds can justly denote. According ɔ his usage principal valencies serve to connect atoms, while ıbordinate valencies, usually represented by dotted lines, ɔnnect molecules. These account for the rest of the attraction nd are limited by the amount of space at the disposal of the ıbstituents. The future influence of these ideas upon the ɔience is difficult to forecast. Werner has himself worked ut a theory of ammonium salts which is certainly far from sat-ıfactory in its present form. Largely in consequence of his rork, however, the ideas of partial, split, and subordinate valence re now frequently applied in both organic and inorganic chem-ıtry, especially in formulating molecular compounds; and while here is as yet little consistency in their use, the idea is clearly estined to exert a considerable influence.

The Rare Gases of the Atmosphere.—One of the most famous f modern researches in inorganic chemistry was that carried out y Lord Rayleigh and Sir William Ramsay which resulted in he discovery of several hitherto unrecognized components of the tmosphere. Nearly a hundred years before, Cavendish had, ndeed, subjected a mixture of air with an excess of oxygen to the rolonged action of electric sparks, and, after removal of the roducts of reaction and the excess of oxygen, had always found residue which was not reduced in volume by further treatment f the same kind. This residue he estimated at about 1/120 part y volume of the air originally employed. No one made any ıse of this observation, probably because later investigators ıad no real idea of how accurately Cavendish had worked, so hat no question was raised as to the nature of the residue until ord Rayleigh in 1893 called attention to the fact that nitrogen repared from the air by removing the other known constituents s heavier than nitrogen prepared chemically in the laboratory, y about one part in two hundred—this discrepancy amounting o fifty times the experimental error of the determination. our explanations suggested themselves. The atmospheric ıitrogen might contain oxygen; the 'chemical' nitrogen (from ammonia) might contain hydrogen; the atmospheric product might contain a heavier allotropic form of nitrogen (perhaps N_3) analogous to ozone; or finally it might contain a small quantity of an inert gas of higher specific gravity. The first two possibilities could be easily disposed of by mixing both kinds of nitrogen with oxygen and hydrogen respectively, and again removing these contaminations. The specific gravities of the products were unaffected.

Argon.—Atmospheric nitrogen was then passed over glowing magnesium which absorbed by far the larger part forming a nitride, but left a residue little affected by magnesium, whose specific gravity was now perceptibly higher than that of the original nitrogen. When by repeating the treatment a gas had been obtained nineteen times as heavy as hydrogen, it was introduced into a Plücker tube and its spark spectrum examined. This still showed the lines of nitrogen, but also a spectrum hitherto unknown which furnished convincing evidence that a new element was involved and not an allotropic form of nitrogen. There still remained the remote possibility that the new substance might in some way owe its formation to the processes devised for its isolation, and to settle this question atmospheric nitrogen was passed through a long series of porous clay pipes surrounded by a vacuum. The portion which diffused through the clay was found distinctly less dense than that which remained behind, showing that a heavier component must be present in the original air. Meantime two processes were worked out for preparing the new gas on a comparatively large scale. One consisted in passing atmospheric oxygen over magnesium mixed with lime, the other was an application of the principle of Cavendish in which the nitrogen was oxidized and the product absorbed. In this way it proved possible to obtain several liters of the new gas which was free from nitrogen and possessed the density 39.88 (0 = 32). To this gas the discoverers gave the name *argon*, 'idle,' on account of its hitherto unexampled lack of chemical affinity. Nitrogen had hitherto been considered an inert gas, but this substance proved absolutely incapable of entering into chemical combination. Ramsay writes on this point:

"The methods employed to prepare argon free from nitrogen—namely, by exposing the mixed gases to the action of oxygen in a discharge of electric sparks, and by passing them over red-hot magnesium—show that it cannot be induced to combine with one of the most electro-negative of elements—oxygen, and one of the most positive—magnesium. It also refuses to combine with hydrogen or with chlorine when sparked with these gases; nor is it absorbed or altered in volume by passage through a red-hot tube along with the vapors of phosphorus, sulphur, tellurium, or sodium. Red-hot caustic soda, or a red-hot mixture of soda and lime, which attacks the exceedingly refractory metal platinum, was without action on argon. The combined influence of oxygen and an alkali in the shape of fused potassium nitrate or red-hot peroxide of sodium was also without effect. Gold would, however, have resisted such action, but would have been attacked by the next agent tried, viz., persulphide of sodium and calcium. This mixture was exposed at a red heat to a current of argon, again without result. Nascent chlorine, or chlorine in the moment of liberation, obtained from a mixture of nitric and hydrochloric acids, and from permanganate of potassium and hydrochloric acid, was without action. A mixture of argon with fluorine, the most active of all the elements, was exposed to a rain of electric sparks by M. Moissan, the distinguished chemist who first succeeded in preparing large quantities of fluorine in a pure state, without his observing any sign of chemical combination.

"An attempt was also made to cause argon to combine with carbon by making an electric arc between two rods of carbon in an atmosphere of argon. It was at first believed that combination had taken place, for expansion occurred, the final volume of gas being larger than the volume taken; but subsequent experiments have shown that the expansion was due to the formation of some oxide of carbon from the oxygen adhering to the carbon rods. On absorption of this oxide by the usual absorbent, a mixture of cuprous chloride and ammonia, the argon was recovered unchanged.

* * * * * * * *

"Professor Ramsay has also made experiments on the action of a silent electric discharge upon a mixture of argon with the vapor of carbon tetrachloride; the latter decomposes, giving, not a resin, but crystals of hexachlorobenzene and free chlorine; but the volume of the argon was unchanged. It was all recovered without loss. Next the rare elements titanium and uranium have been heated to redness in a current of argon, with no alteration or absorption of the gas. And more recently, attempts have been made to cause argon to combine with the very electropositive elements, rubidium and caesium, by volatilizing them in an atmosphere of argon. Numerous experiments, in which electric sparks have been passed through argon cooled with liquid air between poles of every attainable element, have also been made, but without result.

* * * * * * * *

"These failures to produce compounds make it impossible to gain any knowledge regarding the atomic weight of argon from a study of its compounds, for it forms none."

The foregoing gives an excellent idea of the thoroughness with which this investigation was carried out. The difficulty regarding the atomic weight referred to in the last sentence was a very serious one, for the unusual properties of the new element made its position in the periodic table one of extreme interest. The molecular weight of argon was settled by its density as 39.88, but since it forms no compounds the only clue to the number of atoms in the molecule must be sought in physical constants. Fortunately one was available which had already been well studied in the case of known elements and which therefore furnished a reliable analogy. This was the ratio of specific heat at constant volume to that at constant pressure. In the case of all diatomic gases like oxygen, hydrogen and nitrogen this ratio has a value closely approximating 1.4 whereas in monatomic elements (mercury vapor being the most convenient example) its value is 1.6. Furthermore this is not a mere empirical coincidence, since arguments can be derived from the kinetic theory to show why such a difference must exist. The ratio of specific heats for argon was found to be 1.6 and the gas was therefore accepted as monatomic, a conclusion in harmony with its other properties, since a substance which combines with no other element would also be unlikely to combine with itself. The atomic weight therefore is equal to the molecular weight, 39.88, a figure very close to that of potassium.

Terrestrial Helium.—This left argon at first without analogies but it was not destined to remain long unique. In 1868 during an eclipse, the spectrum of the sun's chromosphere displayed certain lines which could not be identified with those of any element known on the earth, and Lockyer and Frankland ascribed them to a new element to which, from the circumstances of its first observation, they gave the name of *helium*. Somewhat later William F. Hillebrand of the U. S. Geological Survey, on heating the mineral cleveite, obtained an indifferent gas which he described as nitrogen, and which therefore received little further attention, until Ramsay in looking for compounds of argon, again prepared it, removed the nitrogen by magnesium, and then observed that the residual gas showed the spectrum of helium. Later the same gas was found in a number of other minerals, notably in monazite sand, where we shall have occasion to discuss its occurrence later, and also in the gases evolved by a number of mineral springs. The new element is, with the exception of hydrogen, the lightest gas known. Its other properties are practically those of argon, and its atomic, like its molecular weight, is 4.

Neon, Krypton and Xenon.—By 1897 the existence of these two gases had convinced Ramsay that a whole new family of elements must exist of which argon and helium represented two members. A thorough search for others was therefore begun and ultimately crowned with success. Numerous minerals were first heated without result, and the waters of mineral springs carefully examined, but these sources at first yielded, in addition to the well-known gases, only argon and helium. Finally the elements sought for were found in extremely small quantities in the atmosphere. The residue from boiling off a large quantity of liquid air was first fractionated and ultimately yielded another inactive gas which was named *neon*, 'new,' while a similar residue from crude argon yielded a still heavier substance of the same type which received the name *krypton*, 'hidden.' A particular interest attaches to this gas because the lines of its spectrum can be observed in the *aurora borealis*. At last especially large quantities of liquid air were employed and helium as well as still another inert gas was found, to which the name *xenon*, 'stranger,' was given. There is no opportunity here to describe the details of this investigation but it should be said in passing that it represented the finest experimental work yet done with gases.

The relative quantities of these new elements which exist in the atmosphere are approximately as follows:

Helium....................	1 part in 245,300	by volume
Neon......................	1 part in 80,800	by volume
Argon.....................	1 part in 106.8	by volume
Krypton...................	1 part in 20 million	by volume
Xenon	1 part in 170 million	by volume

It will be seen that there is a far smaller proportion of xenon in the atmosphere than there is of gold in sea water, and this gives some idea of the skill and patience required to isolate it in a pure state. The research was brought up to this point by 1900 and when the atomic weights of the new elements are compared, it is found that Ramsay's prediction of 1897 is fully justified. They constitute a natural family occupying a period of their own in the table of Mendelejeff where (having a valence of zero) they occupy an appropriate place next to the elements of valence one, and where their neutral properties form an appropriate transition from the most electropositive to the most electronegative elements. A selected portion of the table is here reproduced to emphasize these relations:

Hydrogen 1	Helium 4	Lithium 7	Beryllium 9
Fluorine 19	Neon 20	Sodium 23	Magnesium 24
Chlorine 35.5	Argon 40	Potassium 39	Calcium 40
Bromine 80	Krypton 82	Rubidium 85	Strontium 87
Iodine 127	Xenon 128	Caesium 133	Barium 137

Striking as these results are, the reader will notice that if the elements were arranged strictly in the order of their atomic weights the positions of argon and potassium would have to be reversed. This anomaly is outside the range of experimental error, and while by no means fully explained, it evidently is of the same kind which has long been observed in the relative positions of iodine and tellurium. We shall see later that new light has recently been thrown upon cases of this kind by the work of Mosely upon the atomic numbers.

Literature

There is a *Memorial Lecture* on Victor Meyer in the English collection already so frequently referred to, and an interesting biographical sketch by his brother Richard Meyer which appeared in the *Berichte* for 1908, vol. 41, p. 4505.

Moissan's *Le Four Électrique* appeared in 1897, and his *Le Fluor et ses Composés* in 1900. Gutbier's *Zur Erinnerung an Henri Moissan*, Erlangen, 1908, is devoted almost exclusively to a discussion of his scientific work.

Werner's *Neuere Anschauungen auf dem Gebiete der Anorganischen Chemie*, 3rd edition, Braunschweig, 1913, gives a full exposition of his view of valence as applied to inorganic compounds. The clearest introduction to the subject is, however, still to be found in his article in the *Zeitschrift für Anorganische Chemie* for 1892, vol. 3, p. 267.

Sir William Ramsay's *The Gases of the Atmosphere*, 4th edition, London, 1915, is written in popular language and is extremely interesting.

CHAPTER XX

THE RISE OF PHYSICAL CHEMISTRY

We have seen how frequently chemistry has derived great advantages from the contributions of those who could bring to its problems the equipment and point of view of the trained physicist. Such a service was performed when Boyle and his associates delivered the science from the baleful mysticism and superstition in which alchemy had enveloped it, and again when Lavoisier dispelled the fog with which the vagaries of the phlogiston theory had surrounded the phenomena of combustion. In all these cases the men who ever seek new phenomena had acquired undue influence and had given free rein to their phantasy, so that it was necessary that others who could weigh, measure and define, should control their observations and connect the facts they had discovered by relationships capable of exact mathematical expression.

The rise of modern physical chemistry marks a new movement of this kind which has exercised a dominant influence upon the science since the last decade of the nineteenth century. In one sense physical chemistry is not modern. At all periods since the time of Lavoisier certain eminent investigators have preferred to devote their attention to the borderland between the two sciences. An early instance is that of Berthollet, who at the very beginning of the nineteenth century, attempted to impress upon an inattentive world the important facts that the course of a chemical reaction depends not only upon the affinities involved, but also upon the masses of the reacting substances; that chemical reactions lead to states of equilibrium; and that the physical properties of the products, such as solubility and volatility, exert an important and sometimes a determining influence upon their course. We have seen how, unfortunately for Berthollet, he allowed his reasoning to lead him to conclusions which contradicted the law of definite proportions, with the result that not only were his conclusions discredited but his point of view, so that discussions on similar topics remained unpopular for a long time.

Much of the work of Gay-Lussac as well as of Dulong and of Regnault may be classed as physico-chemical in the best sense, and Bunsen's influence in this direction was preëminent, especially in his work on optics, on the spectroscope, and on the influence of light upon chemical reactions. It is not surprising that he was fond of saying, "*Der Chemiker der kein Physiker ist, ist gar nichts!*" Kopp, the great historian of the science, also did valuable work upon the physical properties of organic compounds as functions of their constitution, which received early recognition on account of its direct application to problems of structure.

Other generalizations, too, which we are in the habit of associating almost exclusively with modern physical chemistry really received attention and were accurately stated at a comparatively early day, at least so far as general principles were concerned. This applies with especial force to the law of mass action, which now ranks as one of the main foundation stones of the science.

The Law of Mass Action.—In 1850 Ludwig Wilhelmy, then a docent at Heidelberg, published a brief paper on the inversion of sugar by acids in which, by means of the polariscope, he studied the progress of hydrolysis with different acids, different quantities of acid, varying temperatures and varying amounts of sugar, and worked out a mathematical expression for the velocity of the reaction, which takes account of these factors completely and correctly. He also pointed out that similar studies upon other reactions of the same type must yield equations of the same form. This proved to be the case, but Wilhelmy himself received little credit, for by the time interest in such problems had become general his work was practically forgotten. We have seen how Berthelot and Péan de St. Gilles about 1860 carried on an important investigation upon the hydrolysis of esters, and expressed their results in the form that at any moment the rate of reaction is proportional to the amount of ester remaining undecomposed. On account of the prestige of Berthelot this work came to more general notice, and about 1863 two Norwegia scientists, Guldberg and Waage, being impressed by the work

Berthelot, gave the idea more general form, and in an extensive investigation they set forth the universal application of the law that at equilibrium the velocity of a chemical reaction is dependent upon the products of the concentrations of the reacting substances. In their principal paper published in 1867 they express this as follows, in reasoning which sounds characteristically modern:

"If we assume that the two substances A and B change by double decomposition into two new ones A′ and B′, and that under the same conditions A′ and B′ can change into A and B, then neither the formation of A′ and B′ nor the re-formation of A and B will be complete; but at the end of the reaction there will always be present the four substances A, B, A′ and B′, and the force which causes the formation of A′ and B′ will be held in equilibrium by that which causes the formation of A and B.

"The force which brings about the formation of A′ and B′ increases proportionally to the affinity coefficient of the reaction

$$A + B = A' + B'$$

but it also depends upon the masses of A and B. We have concluded from our experiments that the force is proportional to the product of the active masses of the two bodies A and B. If we designate the active masses of A and B with p and q, and the affinity coefficient with k, then the force = k.p.q.

* * * * * * * *

"If the active masses of A′ and B′ are p′ and q′ and the affinity coefficient of the reaction

$$A' + B' = A + B$$

equals k′, then the force tending to re-form A and B equals k′.p′.q′.

"This force is in equilibrium with the first and consequently

$$kp.q = k'p'.q'$$

"By experimentally determining the active masses p, q, p′ and q′ the relationship between the affinity coefficients can be found. On the other hand when this relationship is known the result of the reaction can be calculated in advance for any chosen proportion of the four substances at the beginning."

It is not perhaps superfluous to quote the closing paragraph of this important paper:

"Investigations in this field are doubtless more difficult, more tedious ɿd less fruitful than those which now engage the attention of most ıemists, namely the discovery of new compounds. Nevertheless it our opinion that nothing can so soon bring chemistry into the class ith the truly exact sciences as just the line of research with which ıis investigation deals. All our wishes would be fulfilled if we might y this piece of work direct the permanent attention of chemists toward branch of the science which since the beginning of the century has ıquestionably been far more neglected than it deserves."

The Phase Rule.—It was much the same with the phase rule hich, in recent times, has served such an excellent purpose in emonstrating the important relationships involved in heteroeneous equilibrium. In this subject also the essential undering principles were worked out abstractly by Willard Gibbs of ale University as early as 1876. Gibbs, however, was so indifrent to fame that he apparently did not care whether he was so uch as understood by his contemporaries, so that he not only lled no attention to his results, but when it came to publication e buried them in the *Transactions of the Connecticut Academy*. oncerning the importance of the material there concealed stwald has expressed himself as follows:

"To give an idea of the significance of this work it sufficies to say that very considerable part of the laws and relationships which have the meantime been discovered in physical chemistry and which ave led to such an astonishing development of that field within the st decade[1] are found in this paper more or less thoroughly developed. he questions which concern the equilibria of complex systems are here eated with unexampled comprehensiveness and completeness; and addition the influences which are usually considered, such as temperaıre and pressure, there are also discussed the effects of gravity, elascity, surface tension and electricity. Experimental research has only owly begun to follow the paths whose goal and direction are indicated this work, and a wealth of scientific treasures still await experimental eatment, though in many cases this would be an extremely simple atter.

"In the face of such conditions one must ask: Why did this work hieve no success commensurate with its importance? Why, immediely upon its appearance, did not those effects follow which have nce been attained in another way? There are many answers. Above

[1] Written in 1896.

all the blame must be laid to the uninviting form in which the author has recorded his results. In a strictly mathematical manner, and with a text so concentrated that every page requires the active coöperation of the reader, the author takes us through his 700 equations, only seldom illuminating his results with any suggestive applications."

In short this paper by Gibbs must be classed with the *Statique Chimique* of Berthollet which Ostwald himself once characterized as "much praised and little read."

The Theory of Electrolytic Dissociation.—For similar reasons interest in the physical side of chemistry did not become widespread until the theory of electrolytic dissociation was propounded by Arrhenius in 1887. This generalization, which found the world quite unprepared when it was first announced, nevertheless rested upon important chains of evidence which had been in process of development for a long time. These had to do with several widely separated departments of the science, and it was the service of Arrhenius to trace the connection between these facts, and to weld them into a comprehensive whole. It will be well to trace the history of some of these movements in detail.

Hittorf's Work on Electrolysis.—Faraday's law rests upon the fact that whenever a current passes through an electrolyte the latter is decomposed, and for a given quantity of electricity certain definite quantities of the decomposition products appear at the electrodes in chemically equivalent proportions. Faraday rightly concluded that these components of the electrolyte are the carriers of the current and to these carriers he gave the name *ions*. He regarded them as formed by the current and would doubtless have explained the mechanism of the process in the terms of Grotthuss (see page 88) which were universally accepted at that time. Since the quantities of material appearing at the two electrodes are chemically equivalent it was entirely natural, at the time when he wrote, that Grotthuss should make the tacit assumption that both the anion and cation (to use words not current in his day) migrated with equal velocities. This was the universal assumption until, in 1853, Hittorf began a remarkable series of investigations in which he showed that this was not the case. He pointed out that if they wander with different velocities the fact must be susceptible of experimental proof by llowing the changes in concentration which take place about the ectrodes. He said:

"If the two ions do not move with equal velocity, if they do not meet the middle, then that side of the liquid where the more rapidly oving ion makes its appearance will be enriched by a half equivalent that ion, and impoverished by the loss of half an equivalent of the her. The illustration shows this for the assumption that the anion averses one-third and the cation two-thirds of the distance. The le of the liquid at the anode contains after the decomposition one-ird of an equivalent of the anion more, and two-thirds of an equiva-nt of cation less than before. The other side shows the converse lationship."

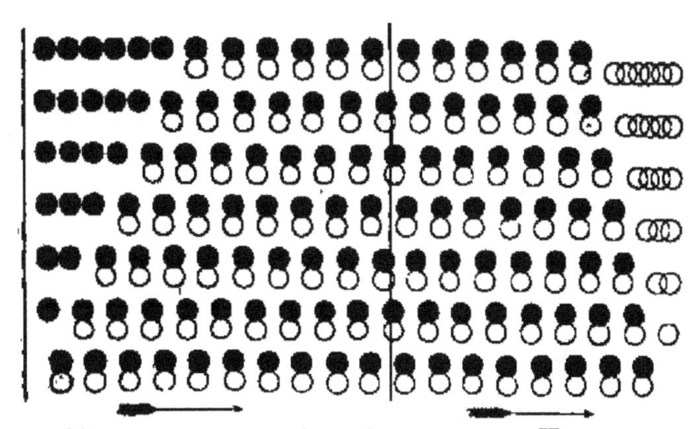

MIGRATION OF THE IONS ACCORDING TO HITTORF

Hittorf proceeded to test his conclusions by the electrolysis of great number of salts under conditions which avoided mechan-l mixing of the solutions and found that the ion velocities were a rule unequal. To this ratio of the velocity of one ion in ms of the other he gave the name of *transference number.* addition to this valuable experimental work Hittorf added portant general discussions in which he pointed out that the e of decomposition of electrolytes by the current is something ich stands in no relationship to the heats of formation of the substances from their elements; that it takes place, in general, ng the same lines of cleavage and into the same components ich interact in double decomposition, and this he endeavored to emphasize by promulgating the dogma, "Electrolytes are salts." Hittorf's conclusions assumed of course that it was the electrolyte and not the solvent which carries the current, an assumption by no means universally shared by his contemporaries. If we follow out this idea we see that in a dilute solution, when an ion has once been set free, it must be at a greater distance from another ion with which it may unite than from many molecules of the solvent, that is, it must be free during most of its transit. Hittorf did not emphasize this conclusion, but this made little difference historically, for his work, excellent as it was, was ignored on the experimental side, and violently attacked on the theoretical one, so that it received no fitting recognition till many years later. Meantime certain theoretical speculations of Clausius began to exert some influence in the same direction.

Views of Clausius.—In 1851 Williamson in a discussion of the mechanism of ether formation had advanced the idea that in any chemical system atoms and molecules must always be in a kind of dynamic equilibrium, and that a molecule, instead of being a rigid structure always made up of the same atoms, is really carrying on a constant exchange with the corresponding atoms of neighboring molecules. Clausius in a paper published in 1857 found this idea useful in explaining the phenomena of electrolysis. He pointed out that if the molecules of an electrolyte were really rigid aggregates then we should expect that with a low potential difference between the electrodes no current would pass. When, however, the electromotive force attained a strength sufficient to disrupt these aggregates a strong current should suddenly result. Experience contradicts these assumptions. As a matter of fact some current passes no matter how low the potential difference, and if this is increased then the strength of the current increases proportionally in accordance with Ohm's law. Clausius felt that these facts could be made more intelligible by the application of an idea like Williamson's.

If we imagine a given potential applied between two electrodes immersed in a solution where a molecular interchange like that described above is going on, then, to quote Clausius:

"A free part-molecule will no longer follow the irregular changing directions toward which it is driven by heat movements, but will alter the direction of its movements in the sense of the force now acting, that among the movements of the positive part-molecules, although ey are still extremely irregular, a certain definite direction will pre- minate, and similarly the negative part-molecules will move princi- lly in the opposite direction. Furthermore, by the action of part- olecules upon whole molecules and of the latter upon each other ose decompositions will be facilitated by which the part-molecules n at the same time follow the electric force in their motions, and ese decompositions will take place more frequently, so that even in ose cases where the position of the molecule is not so favorable that ch a decomposition could take place spontaneously yet the electric rce can cause it to take place. Conversely decompositions by which e part-molecules would be obliged to move against the electric force uld be made more difficult by the force and would take place more dom. * * * It is easy to see that the influence which the electric force erts upon the spontaneous but still irregular decompositions and ovements of the molecules does not first begin when the force has ached a certain strength, but that even the smallest force acts in ch a way as to modify these in the manner explained above, and at the magnitude of this action must increase with the strength the force. The whole process agrees extremely well with Ohm's law."

The Contribution of Kohlrausch.—The next important con- ibution to this subject was made by Friedrich Kohlrausch, who t long after began an extensive series of experiments upon the nductivity of solutions. Only slow progress was made at first cause no really accurate methods were available. When a lution is electrolyzed by a direct current, the polarization which on takes place at the electrodes makes it almost impossible to tain exact values. Ultimately, however, this difficulty was ercome by the use of alternating currents, but it was not till 376 that Kohlrausch published the important paper in which fully confirmed the work of Hittorf. Here he showed that r salts having a common ion, like sodium chloride and potas- um chloride, the conductivities varied inversely as the trans- rence numbers of the common ions in the two salts. Using this ta he proceeded to demonstrate that every ion, regardless of e nature of the salt in which it was apparently combined, had certain definite mobility, or relative migration velocity, which as the same for all combinations, so that the conductivity of a ven salt could be calculated additively from the mobilities of e component ions. These facts were only in harmony with the assumption that during electrolysis the ions were free throughout their course. The modern theory of electrolytic dissociation, however, states that the ions exist ready formed in the solution, even when no current is passing. Kohlrausch did not draw this conclusion, and in order to appreciate the evidence which permitted Arrhenius to draw it later, we must review the historical development of our knowledge concerning certain other properties of solutions.

Raoult on the Freezing-point of Solutions.—The familiar phenomenon that salt water freezes at a lower temperature than fresh first received serious attention from Blagden, an assistant of Cavendish, who studied the matter far enough to learn that, for solutions of the same compound, the depression of the freezing point is proportional to the concentration. Little further work was done until, in 1881, F. M. Raoult, then professor in Grenoble, made an extensive series of experiments along the same lines. These yielded the important additional information that for solutions of different substances in the same solvent the depression of the freezing point was inversely proportional to the molecular weight of the solute, or, as Raoult expressed it: "The quantities of different substances which depress the freezing point by an equal amount are those which the chemist calls molecular quantities." Not long after, Raoult found that an entirely analogous uniformity is to be observed in the elevation of the boiling-point. These observations at once attracted the attention of organic chemists, who had hitherto possessed no reliable method for determining the molecular weights of non-volatile substances, and the technique of the determination as applied to such processes was much improved[1] by Beckmann. The results obtained by Raoult remained empirical, and, what was more serious, they showed numerous exceptions, most of which would be covered by the statement that in aqueous solutions of salts, as well as those of most acids and bases, the observed depressions of the freezing-point (and elevations of the boiling-point) were considerably

[1] Van't Hoff relates the following: "Being in Paris some years ago I asked Raoult's mechanician Baudin to furnish me with a thermometer just like that of Raoult. He, however, strongly advised me against this, remarking: 'The thermometer which Raoult uses is antediluvian!' Nevertheless, with this 'antediluvian thermometer' the world was conquered."

eater than the rule required. The underlying reason for the :ceptions and for the rule itself was at last found by the study : what at first sight seems an entirely distinct class of phenomena.

Osmotic Pressure.—In 1748 Abbé Nollet tried the following ɛperiment. He filled a small cylinder with alcohol, closed the ıouth with a membrane so as not to include any air, and imıersed the whole in water. He was surprised to see that water ntered through the membrane until the latter was much disended, and on piercing it the liquid "spurted about a foot." 'his seems to have been the first scientific observation of the ›henomenon of osmosis, which we now interpret essentially as ollows: In any solution the dissolved substance exerts a pressure ıgainst the surface of the liquid tending to expand it. In an ›rdinary vessel this is balanced by the surface tension and proluces no visible effects. Animal and vegetable membranes, ıowever, have the peculiar property of being permeable for vater but not for most substances dissolved in it. In an ex›eriment like that of Nollet, then, the liquid can expand because he solvent can now enter through the membrane. The force ausing this expansion we call the osmotic pressure, and we know hat it depends upon the solute rather than the solvent because t varies with the nature and concentration of the former. Measırements of the magnitude of this force were at first difficult to ›btain because natural membranes are not absolutely imperneable to the molecules of dissolved substances. Traube, ıowever, by forming colloidal precipitates like copper ferro:yanide in the pores of clay cups, was able to prepare cells which ulfilled these conditions, and by their use in 1877 Pfeffer, then ›rofessor of botany at Bonn, made some quite accurate measurenents of the force. He found it to be of an unexpected magniude, a one per cent. sugar solution, for example, exerting a ›ressure of about two-thirds of an atmosphere. The subject ›f osmotic pressure has always been of exceptional interest to ›hysiologists and botanists because it plays an important rôle vherever there are cellular tissues, and plants owe to it their irculation and growth. In the early eighties Hugo de Vries f Amsterdam was making experiments upon the withering of lants. He found that, when placed in pure water, they showed a ɜndency to swell, whereas solutions more concentrated than those in the plant cell exerted a dehydrating action and the plants withered. Between these extremes it was possible to prepare so-called isotonic solutions which neither gave water to the plant cell nor took water from it; that is, they had the same osmotic pressure as the cell. De Vries prepared such solutions from a considerable number of different salts, and then made the important observation that these isotonic solutions, which he knew had the same osmotic pressure, all had the same freezing point.

The Contribution of Van't Hoff.—In 1884, when the investigation had reached this stage, de Vries almost by accident communicated his results to Van't Hoff, who at once saw their importance for physics and chemistry. He proceeded to work out the causal relationship which exists between osmotic pressure on the one hand, and vapor pressure, boiling-point and freezing-point on the other. He then made an extensive study of the nature of osmotic pressure in general, and found that the relationships involved are far more simple than had been supposed. When a substance is dissolved in a liquid its molecules exert against the surface of the latter a pressure which is not only analogous, but in most cases numerically equal, to the pressure which they would exert if the substance were a gas confined in the same volume. If, therefore, we write the equation of state for a gas:

$$PV = RT$$

where P is the pressure, V the volume, T the absolute temperature, and R the so-called gas constant, the same equation holds for a substance in solution even to the numerical value of R.

There were, however, exceptions, and since the osmotic pressure determines the freezing point, these exceptions were the very ones already observed by Raoult, who had found, as we remember, that most salts and some acids and bases showed a greater depression of the freezing point than he could account for. This meant that their osmotic pressure was abnormally high, a fact which might be interpreted to signify that more molecules were present in the solutions of these substances than existing theories could account for. Van't Hoff had, at the time, no explanation to offer, but contented himself by writing the equation of state for such substances:

$$PV = iRT$$

which *i* represented a constant dependent upon the nature of individual substance. The equation appeared in this form ;he now famous paper on the subject which he presented to the edish Academy of Sciences in 1885.

Early Views of Arrhenius.—Meantime Svante August Ar-nius, then only about twenty-four years old and just complet- his studies at Stockholm, had presented to the same body communication upon his recent work in electrolysis. This tained some ideas which later investigation has shown to be oneous or incomplete, but it may be said in general that he roached the subject essentially from the point of view of usius. He concluded from his experiments that in any ducting solution only a certain proportion of its particles are lly responsible for the conductivity, which in different solu-is will be proportional to the relative quantity of such particles compared to the rest. At this time Arrhenius made no empt to differentiate these particles qualitatively from the er molecules in solution, but he assigned to every conducting ition a so-called *activity coefficient* to represent the proportion ts particles which took part in electrolysis. When, now, the ;tivity coefficients" of Arrhenius were compared with the fficients which Van't Hoff had designated by *i* in his work on iotic pressures the two were found strictly proportional! other words the better a solution conducts the current the more iormally great is its osmotic pressure, or, as we have seen ve, the more molecules it seems to have in a solution of given centration. Clearly this effect would be produced if the lecules were dissociated.

Ostwald on Affinity Constants.—An entirely independent ument which spoke in the same sense was next furnished by wald who had for some time been studying the so-called *nity constants* of organic acids. By this term was meant the ength of the acids as measured by the velocity with which se substances catalyse the hydrolysis of esters or the in-sion of sugar. Ostwald had determined this constant for ty or more organic acids, and now measured their con-tivity as well, and found his "affinity constants" were now portional to the "activity constants" of Arrhenius. This eement is what must be expected if in any solution the ions exist ready-formed, for the more ions are present the greater must be the conductivity, the greater also will be the number of free particles and hence the osmotic pressure, and finally since all acids catalyse hydrolysis, and since the effect must be due to the only thing which all acids have in common—the hydrogen ion—the more ions are present, the greater must be the velocity of the hydrolysis.

Final Statement of the Theory.—On the strength of this unanimity Arrhenius in 1887 promulgated the theory of electrolytic dissociation in essentially its present form, and pointed out, in addition to the arguments already enumerated, that where dissociation is practically complete, as in the case of most salts, the physical properties of their solutions must be additive functions of the corresponding properties of the individual ions. He showed that this is actually the case not only for the properties already mentioned but also for others like specific gravity and volume, refractive index, and capillarity. He also showed how simply the new theory accounted for the hitherto puzzling fact that in dilute solution the heat of neutralization of all acids and bases is the same. If we accept the dissociation theory, the one thing which all these reactions have in common is the formation of undissociated water from the hydrogen ions of the acid and the hydroxyl ions of the base, the other ions remaining free. The heat evolution then in all cases is the heat of this reaction:

$$\dot{K} + OH' + \dot{H} + NO_3' = H_2O + \dot{K} + NO_3'$$

Its Reception.—The new theory impressed most chemists as revolutionary, and while little could be said against the facts upon which it was based yet it had to meet with a great deal of 'passive resistance.' Its acceptance required a new mental attitude. Sodium chloride, for example, had always been considered an extremely stable compound, doubtless because it has a high heat of formation. The new doctrine, however, seemed to teach that we have only to dissolve it in water in order to decompose it into its elements, in spite of the fact that the chlorine reveals itself neither by its odor or its color, and the sodium does not evolve hydrogen with the water. These objections were of course chiefly the result of misunderstanding, since the chlorine for example, is not the element, but an atom thereof plus a ;ain definite charge of electricity. In order to overcome such understandings the three investigators most concerned in the ablishment of the new theory, Ostwald, Arrhenius, and Van't ff, formed a kind of offensive and defensive alliance for the pagation of the new faith. Perhaps the largest factors in the cess achieved by the coalition were the foundation of the *tschrift für Physikalische Chemie* by Ostwald and Van't Hoff 1887, and the teaching and literary activity of the former ing the next twenty years.

)stwald.—Wilhelm Ostwald was born in Riga August 21, 1853, l entered the local *Realschule* in 1864. Here he required seven rs to complete the course to which most pupils devoted only , but there is abundant evidence to show that this was not to any lack of capacity, but to an altogether unusual versa- y, and a tendency to follow many lines of intellectual activity self-instruction outside the curriculum. This manifested lf in the collection of insects, the manufacture of fire-works, teur photography (which then involved the difficult man- ations of the old 'wet process'), carpentry, bookbinding, iting, and the equipment of a private laboratory where he ld pursue chemical work beyond the point attainable in ool.

1 1871 Ostwald entered the University of Dorpat where he at devoted himself mainly to the frivolities of student life, but r settled seriously to work, and under the stimulus of a ernal warning took his degree within a space of time which e of his acquaintances then believed possible. In the same r, 1875, he became assistant in physics at Dorpat, a position ch he retained till called to a professorship at Riga in 1881. ile there his physico-chemical work, especially upon the iity constants of acids, became widely known, and in 1887 he made professor of physical chemistry at Leipzig. This date a significance in the history of chemistry akin to that when oig was called to Giessen, for the work of Ostwald, Arrhenius, Van't Hoff was now attracting wide attention, and it served only to win adherents for the theory of electrolytic dis , but also to arouse a latent interest in students th r, who now began to realize that they wished to stu

istry from the physical point of view. These now flocked to Ostwald's laboratory in Leipzig, which became, as had been said of that at Giessen fifty years before, the factory for producing the world's supply of professors of physical chemistry. Accounts of those who studied there in the early days agree that the atmosphere was most inspiring. Laboratory conveniences were at first of the most meager description, but professor and students lived, as it were, together, all ideas were shared in common, and the field being practically new, important discoveries followed each other in quick succession, Ostwald himself being an example of tireless energy to all. In addition to the routine work of teaching, the supervision of research, the publication of results and the editorial duties of the *Zeitschrift*, he has found time to write many books, including the large *Lehrbuch* in six volumes, the *Elektrochemie* (a monumental historical study), numerous books illustrating different methods of teaching chemistry, and others dealing with philosophy, biography and painting. He also has been a prominent agitator for reform in the German schools, as well as for the introduction of an international language, and he is a painter of acknowledged merit. It is small wonder that such a man should call his country house "*Energie.*"

Ostwald's voluminous writings have extended his influence far beyond the walls of his laboratory, and thus helped many a teacher to improve his own methods of introducing students to chemistry—a matter in which the author takes pleasure in acknowledging his personal indebtedness. In one point, however, Ostwald has worked against the tendency of the times. In his anxiety to remove from chemistry all superfluous hypothetical elements, he has systematically discouraged the kinetic point of view, and has taught chemists to avoid all explanations of natural phenomena which involve the assumption of discrete particles, even the atoms of Dalton. Here, also, his influence has doubtless been a valuable corrective, but since the beginning of the present century evidence has been constantly accumulating to prove the objective reality of the atom and the discontinuity of even important forms of energy like light and electricity, so that modern physics, which was practically free from corpuscular speculations when Ostwald began his campaign, has now gone over almost completely to that point of view.

OSTWALD AND VAN'T HOFF

OSTWALD AND ARRHENIUS

(Facing page 248)

HENRI BECQUEREL
1852–1908
Reproduced from a photograph by the kind permission of Ch. Gerschel, Paris.

MADAME SKLADOWSKA CURIE

Other Contributions of Physical Chemistry.—To students of the present day it is altogether superfluous to attempt any detailed account of the service which physical chemistry has done to the science as a whole. The ideas it has introduced now permeate all instruction in chemistry and make their influence felt in every department of the science. To inorganic chemistry, especially, has come the inspiration of a new point of view and a re-awakened enthusiasm toward research. Dealing, as it does, with the influence upon chemical reactions of temperature, pressure, concentration and catalysis, physical chemistry has given a new insight into the mechanism of all chemical change, and made it possible to fix, as never before, the conditions most suitable for a given effect. There has resulted not only an enhanced accuracy in analytical procedure but a universal improvement in laboratory technique. Industrial processes also have benefited universally, for physico-chemical reasoning has in many cases made it possible to calculate in advance the most economical conditions of their operation. The contact process for the preparation of sulphuric acid is a beautiful example of this, but they might be multiplied indefinitely. Organic chemistry, alone, has as yet derived the least benefit, because so many of its reactions are complicated by tarry by-products and by side-reactions which make difficult any attempt to make accurate measurements. Real advantage is to be hoped for, however, even here, because in the past physical chemistry has established some of its most important generalizations in the study of organic material. Examples are to be found in the reversibility of chemical reactions, the mass action law, and the laws of dissociation as exemplified by weak acids.

Origin of the Galvanic Current.—Before leaving this topic it may be appropriate to complete in a few words the story of the long controversy concerning the origin of the current in the galvanic battery. It will be recalled that Volta had disregarded the chemical phenomena associated with the operation of his *pile*, and ascribed the origin of the current to the potential difference between the metallic plates. He demonstrated this by joining the dry plates, separating them and examining each with an electrometer. They then showed a difference of charge to which he attributed the current, and on this basis he constructed his potential series. Not long after, others, to whom the chemical action of the current especially appealed, pointed out that chemical action within the cell must furnish the energy for whatever was accomplished outside it by the current, that the two were proportional, and hence they must be related as cause and effect. Faraday's law established the first of these propositions beyond question, but the 'physicists' were unwilling to concede the last, because the repetition of Volta's original experiment with the electrometer always gave a difference of potential between the dry plates. As these experiments were repeated with more and more care it was found that the potentials observed were extremely dependent upon the surface condition of the plates, but no valuable conclusion was drawn from this. About 1870, first Le Roux and then Edelund, using thermo-electrical measurements on welded metallic contacts, showed unmistakably that the potential differences existing between dry metals are far smaller than Volta had supposed, and not even of an order of magnitude which could account for the action in a voltaic cell. This made it clear that since the metals alone were not responsible for the current, it must originate at the contact between metal and liquid, but here again we have to ask at which contact. In a Daniell cell, for example, zinc is going into solution and copper is being precipitated. The cell only operates when both processes are going on, so that one cannot be isolated and measured. Ultimately this difficulty was also overcome. In 1885 Helmholtz, basing his arguments upon some interesting, but at first sight entirely irrelevant, experiments upon the reciprocal relations of potential and surface tension in metallic mercury, came to the conclusion that when a rapidly dropping and isolated mass of mercury is in contact with an electrolyte at the dropping point, the electrolyte and the mercury can exhibit no difference of potential. "A dropping mass of mercury therefore forms," as Ostwald points out," an electrode by which one can connect liquids with an electrometer without change of potential." It was suitable therefore for measuring the other potential differences in the cell. By this means the chemical origin of the current was established by Ostwald in 1887.

Literature

Many of the papers referred to in this chapter are republished in OSTWALD'S *Klassiker*. That by WILHELMY is in No. 29, that by BERTHELOT and PÉAN DE ST. GILLES in 173, that by GULDBERG and WAAGE in 104, that by BLAGDEN in 56, that by VAN'T HOFF in 110, that by HITTORF in 21 and 23, and that by ARRHENIUS in 160.

COHEN'S life of Van't Hoff, already alluded to, gives an interesting account of how the latter became interested in osmotic pressure, while OSTWALD'S *Elektrochemie* sets forth with admirable clearness the historical development of the dissociation theory.

WALDEN'S *Wilhelm Ostwald* is as yet the only biography. Many of Ostwald's own writings have been repeatedly referred to.

CHAPTER XXI

RADIOACTIVITY—ITS INFLUENCE UPON THE ATOMIC THEORY

The last years of the nineteenth century witnessed some remarkable developments in the science of physics which have deeply affected many fundamental conceptions hitherto considered within the province of chemistry, such as the nature of the atom and the ultimate composition of matter. These developments were associated with the discovery and study of certain new and altogether unusual types of radiation.

The X-rays.—As early as 1879 Sir William Crookes had passed electric currents of high potential through tubes containing gases at exceedingly low pressures—so-called vacuum tubes—and had observed that under these circumstances rays are emitted from the negative electrode or cathode which differ markedly from any hitherto studied. They proceed in straight lines from the cathode, but show the remarkable property of being deflected by a magnet, which would seem to indicate that they represent a stream of minute material particles. These phenomena remained isolated for some time. In 1895 Röntgen found that when the cathode rays impinge upon a solid a new kind of ray is generated, which now penetrates the glass walls and proceeds into space, producing effects then altogether novel but now recognized as common to most of these new types of radiation. They cause, for example, various substances like zincblende, willemite, barium platinocyanide, etc., to fluoresce, they affect the photographic plate in a manner similar to the action of light, they traverse opaque media, and they ionize gases. By this is meant that gases through which these rays pass become conductors of electricity, so that if such gases are introduced into an electroscope the latter loses its charge, and the relative velocity with which this occurs may be used as a measure of the relative ionizing effect of different types or sources of rays.

Radioactivity.—This new form of radiation received the name of X-rays and the fact that indirectly they made it possible to 'see through' objects hitherto considered opaque, excited the greatest popular interest. No one, however, realized that their discovery had any important bearing upon chemical problems until in the following year Henri Becquerel discovered a somewhat similar type of radiation which proceeded from an entirely different source. Becquerel came of a family long eminent for its contributions to the study of fluorescence, and had himself lived up to the family tradition. The formation of X-rays from cathode rays in the vacuum tube and the fluorescence of the glass always observed in the latter suggested to Becquerel that fluorescent substances might possess the property of making a similar transformation of light waves. His first experiments served to strengthen that belief. A uranium salt was exposed to the sun's rays while resting upon a photographic plate wrapped in black paper to protect it from the light. When the plate was afterward developed, the portion under the uranium salt was found to have been affected. On one occasion, however, when all other conditions had been the same, an accident prevented the exposure of the uranium salt to light. Nevertheless the plate exhibited the same effects as before, showing that uranium salts continuously emit rays capable of affecting the photographic plate even when not exposed to light. This fact was confirmed by experiments upon a great variety of uranium compounds. All showed the action, and its intensity was found proportional to the percentage of uranium in the substance. A puzzling exception was observed in the case of pitchblende from Joachimsthal—the mineral which had hitherto served as the chief source of uranium preparations. This mineral exhibited an activity several times greater than that of metallic uranium, showing that it must contain some other substance more highly radioactive than the latter. Madame Skladowska Curie, at the suggestion of Becquerel, now undertook the chemical examination of the pitchblende (which is very complex) in order to find this especially active component.

The Discovery of Radium.—The first material which showed the property in a higher degree than uranium was a substance resembling bismuth, to which Madame Curie gave the name of *polonium* in honor of her native country. It has not even yet been obtained in a state of purity. A little later she discovered among the alkaline earths a new element which possessed the property of radioactivity in an especially high degree. It closely resembled barium but could be separated with some difficulty from the latter by the fractional crystallization of certain salts. To this new element was given the highly appropriate name of *radium*, and its purification was finally pushed to a point which justified a determination of the atomic weight. This yielded the value 226 and entitled the new element to a vacant space in the periodic table just below barium, which is also in harmony with its spectrum and chemical properties.

The properties of radium are remarkable. Its salts are self-luminous, and constantly give off radiations producing effects similar to those of X-rays. It was found that these rays were given off continuously with undiminished energy, and not only were rays emitted but heat—a gram of radium evolving 133 calories per hour. When the character of the radiation was more thoroughly studied it was found that three distinct classes of rays could be distinguished. These are still spoken of as the α, β and γ rays. Of these the α rays are the least penetrating and have least action upon a photographic plate, while, on the other hand, they are the most potent in ionizing gases. The γ rays are the most penetrating and in their character most resemble X-rays. The β rays, however, are identical with, or closely allied to, the cathode rays of the Crookes tube, though they have a higher velocity. They carry a negative charge and are deflected by a magnetic field. The γ rays are not deflected and the α rays but slightly. In the latter case, however, the deflection is in the opposite direction, showing that these rays carry a positive charge. Their essential nature will be discussed later.

Rutherford's Work on Thorium.—The subject of radioactivity entered upon a new stage when, in 1900, Sir Ernest Rutherford, then professor in Montreal, began an intensive study of the radio-activity of thorium, the only previously well-known element except uranium which had thus far shown the property. Acting upon the observation of Owens that some of the effects produced by radioactive products were modified by currents of air, Ruther-

Ernest Rutherford
1871–

(Facing page 254)

d found that air which had been passed over an active thorium paration had itself acquired activity, but that this activity ayed rapidly with the time in accordance with the equation:

$$\frac{I_t}{I_0} = e^{-\lambda t}$$

ere I_0 represents the initial intensity, I_t that at the time t, e the e of natural logarithms, and λ a constant characteristic of the stance. It has since been found that all radioactive materials ow this law of gradual decay and it is perhaps the most im-tant single generalization in the subject of radioactivity. The eriments showed that thorium was continually producing an remely attenuated but highly radioactive gas, and to this therford applied the name of the thorium *emanation.* found that when the gas is retained for some time in any sel the walls of the latter become coated with active material. is he called the *radioactive deposit.* It exhibited some narkable properties. If a negatively charged wire was sus-nded in a vessel containing the emanation, all the deposit was ncentrated upon the wire. The quantity of material was so all that it could be recognized only by its activity, but that was a solid adhering to the wire seemed amply proved by the t that it could be driven off by heat, or removed by rubbing h sandpaper. Still another substance, therefore, had been med by the decomposition of the emanation, whose activity d period of decay were different from the latter. In 1902 therford and Soddy pointed out another decomposition of orium compounds of a somewhat analogous character. When solution of thorium nitrate, for example, is precipitated by monia, the hydroxide thrown down, when filtered and dried, found to be almost inactive. If the filtrate, however, be aporated to dryness and the ammonium nitrate expelled by at, an extremely small residue is left which possesses practically e whole activity of the original preparation. At the end of out a month, however, this has been practically lost, while that the precipitated thoria has by this time regained practically its original value. Subsequent investigation has shown that e changes just mentioned are in reality a good deal more com-x, but these experiments sufficed to prove that radioactivity is accompanied by the formation of new material. Any single process, therefore, cannot be of infinite duration.

The Theory of Atomic Disintegration.—On the strength of this evidence Rutherford and Soddy in 1902 advanced their theory of atomic disintegration, which thus far has accounted for all observed phenomena and is the present working hypothesis of the subject. Its fundamental principles may be stated as follows: The chemical atom is not to be regarded as an impenetrable and indivisible point, but as an extremely complex structure, and the forces which determine the relations of its component parts are incomparably greater than any which obtain in chemical combination between the atoms. The atoms of a substance which we call radioactive are unstable, and manifest this instability in the peculiar manner that certain ones (determined solely by the total number present) decompose explosively every instant, throwing off with great velocity the material composing the various kinds of rays above described, and leaving behind a new chemical element with properties of its own, which may or may not include radioactivity.

Formation of Helium from Radium.—Since radium occurs only in minerals containing uranium, this theory made it probable that radium is a product of the latter, and since most active minerals such as monazite sand contain helium, this might be looked upon as one of the products of such activity. The latter point was established beyond question in 1903 when Ramsay and Soddy undertook a thorough study of the radium emanation. This material is a true gas obeying Boyle's law. It can be separated from other gases, condensed to a liquid, and frozen. Its atomic weight as determined from its density by Ramsay is 222. He classified it in the argon group and named it *niton*. The most striking observation made by Ramsay and Soddy was that when this gas has been kept for some days it disappears and helium appears in its place. This discovery, which represented the first known production of one element from another, seemed a realization of the dreams of the alchemists, and aroused a popular interest almost equal to that excited by the discovery of the X-rays. The experiment was soon successfully repeated in several laboratories.

In the same year Rutherford pointed out that helium could

not well be the only decomposition product of radium, as indeed was improbable, because the radium emanation like that of thorium also yielded an *active deposit*, which probably represents the greater portion of the products of decomposition. Rutherford suggested that the α rays emitted by the emanation, as well as other radioactive substances might really consist of electrically charged atoms of helium. In 1909 he was able to prove this by an extremely ingenious experiment. Some of the radium emanation was sealed into a tube of glass so extremely thin that α rays could penetrate it with considerable ease. This tube was then placed inside another which was attached to a spectrum tube. The outer tube was then evacuated. After two days its contents showed the principal lines in the spectrum of helium and after six the spectrum was complete. As a control a similar experiment was tried in which the inner tube was filled with helium instead of the emanation. None of this, however, penetrated to the outer tube.

Meantime it had been shown that many other radioactive changes take place with evolution of α rays, and it follows from this, that the production of helium is not a particular property of the radium emanation but is a frequent accompaniment of perhaps the majority of such changes.

Effects Produced by Single Atoms.—For our general conception of the nature of matter it is, perhaps, of still greater interest to know that the evolution of helium in such changes is discontinuous, and consists in the expulsion of discrete particles. It has been found possible to prove this by direct observation. α Rays cause the gases through which they pass to conduct electricity, and Rutherford and Geiger succeeded in devising an "ionizing chamber"[1] in which it was possible, when high voltages were employed, to detect the slightest currents by the movement of an electrometer needle. When now a small portion of the α radiation of a weakly active preparation was allowed to enter the chamber, the entrance of each particle gave rise to a ballistic throw of the needle, so that the number of particles entering in a given time could be accurately counted. Had the radiation

[1] An ionizing chamber consists of an enclosed space between t[illegible] plates charged at a high potential difference. No curren[illegible] unless the air between the plates becomes ionised.

represented a continuous stream, the needle would, of course, have shown only a constant deflection.

Essentially the same results were obtained in another way. When radium was first discovered, Sir William Crookes found that the luminescence which it produces upon a screen of crystalline zinc sulphide is really made up of scintillations, apparently caused by particles expelled from the radium striking the crystals, and this led him to construct the familiar piece of apparatus known as the *spinthariscope.* If now in an experiment like that described above we replace the ionizing chamber by such a zinc sulphide screen or by a diamond, it is possible, with the aid of a microscope, to count the scintillations in a given time. The two methods give very concordant results. From either, the number of particles evolved from any preparation in a given time can be calculated. If now we know the total charge carried by the rays during the same time, we have the data for determining the charge carried by each particle, an important constant in radioactivity work.

Mass and Dimensions of the Atom.—Some years before these determinations were made, J. J. Thomson, C. T. R. Wilson and others had made extensive studies of the conduction of electricity through gases in the Crookes tube and similar forms of apparatus, and had come to the conclusion that both the positive and negative current are carried by minute particles, and that both sets of particles carry a charge of equal magnitude (the same which is carried by the hydrogen ion in electrolysis), but that the masses of the two kinds of particles differ widely, those carrying a positive charge being of atomic magnitude, while those carrying the negative current (in the cathode ray) have a mass only $1/1000$ of that of the hydrogen atom. We now call the latter particles *electrons.* Applying the same reasoning to the results of their own experiments Rutherford and Geiger concluded that the α rays expelled by radioactive material consist of atoms of helium, each of which bears two unit charges of electricity. Making this assumption they were able to predict with accuracy the volume of helium which should be evolved by a gram of radium in a year, and thus to furnish an extremely simple and convincing proof of the relationships assumed. Rutherford writes:

"The determination of the number of α particles emitted by radium and of the value of the unit charge allows us at once to deduce values of a number of important atomic and radioactive magnitudes. If e be the unit of charge carried by the hydrogen atom in electrolysis, and n the number of atoms in one gram, it is known from Faraday's experiments that $ne = 9647$ electromagnetic units. Since $e = 4.65 \times 10^{-10}$ electrostatic units, or 1.55×10^{-20} electromagnetic units, the value of n is at once determined. From this it is a simple matter, assuming Avogadro's law, to deduce the number of molecules in one cubic centimetre of any gas at standard pressure and temperature. For convenience some of the more important atomic and radioactive constants are tabulated below:

Charge carried by the hydrogen atom.........	4.65×10^{-10} e.s. units
Value of e/m for α particle...................	5070 e.m. units
Charge carried by the α particle..............	9.3×10^{-10} e.s. units
Number of atoms in one gram hydrogen......	6.2×10^{23}
Mass of an atom of hydrogen.................	1.6×10^{-24} gram
Number of molecules per cubic centimetre of any gas of standard pressure and temperature	2.78×10^{19}

"With the aid of these data, it is possible to deduce at once the rate of production of helium from any substance, for example radium, for which the number of particles emitted per second has been determined. It is known that one gram of radium in equilibrium produces 13.6×10^{10} atoms of helium per second. Dividing by the number of atoms of helium in one cubic centimetre, this corresponds to a production of helium of 4.90×10^{-9} cc. per second or 158 cubic mm. per year. It will be seen that this calculated value is in close agreement with that determined experimentally. Such a close concordance between calculation and experiment affords strong evidence of the essential correctness of the data on which the calculations are based."

Structure of the Atom.—The force of this reasoning as an argument for the objective reality of the atom may perhaps be better appreciated if we abandon for the moment the historical point of view, and consider the question as a problem to be solved by the experimental data now available. The quantities susceptible of direct measurement are the number of particles evolved by a given weight in a given time, the total charge carried by a given number of particles, the total quantity of electricity transported by the radiation of a given amount of radium in a given time, and finally the total quantity of helium evolved. From these data a simple calculation gives the number of particles present in a cubic centimeter of helium. There remains only the question whether these particles are really the chemical atoms, as we have hitherto understood that term. When we find that the number, dimensions, and charges carried by such particles agree entirely with the figures obtained by physicists using more indirect methods, there can no longer be much doubt as to the reality of the atom and probably of the electron. This certainty has naturally stimulated speculation concerning the structure of the atom itself. The phenomena of radioactivity would seem to indicate that electrons and atoms of helium are its most important components, but others may possibly be involved as well. There is as yet very little agreement between individual theorists in such matters. Physicists seem inclined to think of the atom as a highly dynamic complex, analogous to a planetary system, while chemists for the most part favor some conception which will permit close packing of the atoms in the molecules of solids.

The electron, too, is assuming a more and more important place in the chemical vocabulary. It not only appears in modern explanations of electrolysis and of metallic conduction, but chemical combination is frequently interpreted as the transfer of one or more electrons from one atom to another, the residual force which still binds the wandering electron to its original atom acting as a bond of valence. This leads naturally to electronic conceptions of valence in general. Several hypotheses of this kind have been suggested which doubtless contain elements of truth, but in organic chemistry where the valence theory is most important they have as yet hardly demonstrated their usefulness.

Radioactivity and Cosmogony.—Another topic which is of great general interest, but which hardly pertains to our subject, is the influence which the discovery of radium has had upon the fundamental conceptions of geology. The amount of heat which is continually evolved by radium is enormous in proportion to its bulk, that emitted by a cubic centimeter of the emanation in the course of its complete transformation being approximately ten million times as great as that evolved in the combination of the same volume of oxygen and hydrogen. Radioactive materials are, however, widely distributed in the earth's crust, and probably in its interior, so that the total heat which they evolve must compensate in large measure for that continually lost by radiation. In consequence our previous conceptions concerning the earth's period of cooling must certainly be revised. It is easy to see that much more time can now be allowed for the processes of plant and animal evolution, as well as for other geological transformations. Furthermore the rate of transformation of radioactive elements when in equilibrium with each other is now so accurately known, that the minimum age of an active mineral can be calculated from its composition with a good deal of certainty. In some cases the results are as high as sixteen hundred millions of years.

The Products of Radioactive Disintegration.—From the chemical standpoint there is greater interest in certain researches carried on simultaneously with those described above, and which had for their object the discovery of new kinds of radioactive material. These for the most part had to be sought in the transformations of elements already known. Here it was soon found that many products at first deemed homogeneous like the *active deposit* really represented mixtures of several substances passing consecutively into each other. The relationships involved seemed at first sight hopelessly complex, but by applying the law of decay it became possible to recognize the activities of different elements even when they were superposed. In this way elements were discovered whose "period of average life" had sometimes to be reckoned in seconds. This constant varies widely. For radium it is 2,440 years, for the emanation 5.55 days, and for uranium it is millions of years. In all about thirty radioactive elements have been discovered, each of which belongs to one of three series or families. The first comprises the products of disintegration of uranium and contains radium and polonium. The second series is formed by the decomposition of thorium, and the third is derived in the same way from actinium, a natural radioactive element allied to lanthanum, which was discovered by Debierne in 1899. There is some reason to suppose that all three of these series end in ordinary lead, but this has not yet been proved.

The Mechanism of Radioactive Change.—Especial interest attaches to the mechanism of the process by which one of these elements is formed from another. We have seen that radioactive change is almost always accompanied by the evolution of three kinds of rays. Of these we know that the α-rays are charged atoms of helium. Physicists who have made accurate comparison also tell us that the β rays are identical with cathode rays and consist of electrons, while the γ rays are

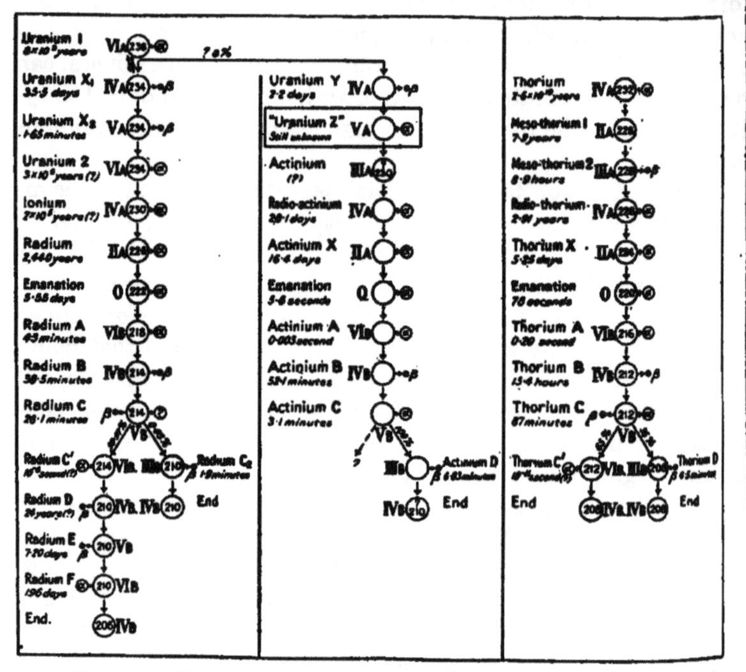

GENETIC RELATIONSHIPS OF THE RADIOACTIVE ELEMENTS

Reproduced from Soddy's "Chemistry of the Radio-Elements" by the kind permission of the author and Longmans Green and Co.

probably of secondary origin, being produced whenever α or β rays impinge upon matter. They are similar to X-rays and contain no material particles. An atom, then, loses either an atom of helium or an electron, according to whether in a given change it emits an α or a β particle, and it seems to be the rule that in any single transformation only one such loss occurs. The resulting uniformity is extremely simple. When an element loses n atom of helium it must of course lose four units of atomic veight. It also loses two units of valence, and the new element vhich results occupies a position in the periodic table preceding y two that of the parent element. This is admirably illustrated y radium. This element belongs to the alkaline earth group and s closely analogous to barium, having an atomic weight of 26 and the valence two. It emits α-rays and goes over to he emanation. This has the atomic weight 222. It has the alence zero and finds an appropriate place in the periodic able in a vacant place below xenon, in the family of the inert ases. If, on the other hand, the original element loses an lectron (β radiation) no change of atomic weight is involved, ut the properties of the resulting element entitle it to a poition in the table one point beyond that of the parent.

III B Thallium	IVB Lead	VB Bismuth.	VIB Polonium	VII B [Iodine]	O [Xenon.]	I A [Caesium]	II A Radium.	III A Actinium	IV A Thorium.	VA [Tantalum]	VI A Uranium.
81	82	83	84	85	86	87	88	89	90	91	92

α-ray change indicated ←——— ——→ Indicates β-ray (or rayless) change.

RELATIONS OF THE ISOTOPES

Reproduced from Soddy's "Chemistry of the Radio-Elements" by the kind permission of the author and Longmans Green and Co.

The Isotopes.—It will occur to the reader that any general application of these laws must result in assigning to some of the newer radioactive elements positions in the periodic table which are already occupied. Marvelous as it seems on the basis of our previous conceptions, this does not make the slightest difference! In spite of the distinguishing property of radioactivity, and in spite of a difference in atomic weight which may in some cases amount to several units, the two elements are alike in all other respects. This does not mean similarity but identity. They are alike in all their properties physical and chemical, except radioactivity, and where sufficient quantities are obtainable to make the experiment, it has been shown that when they are once mixed no separation is possible. Substances which stand in this relation are said to have the same *atomic number* and are called *isotopes*. Furthermore there seems to be no necessary limit to their number. Ordinary lead appears to have no less than four.

The simplicity of the relationships above described has only recently been realized, having been set forth in the work of Fleck, Russell, and Fajans in 1913. By their aid it is obvious that the properties of any radioactive element, no matter how short-lived, can be readily foretold if we know the position of its parent in the periodic table, and the kind of radiation by which its formation is accompanied. This determines the position of the new element in the periodic system. If this position is already occupied, the properties of the new element will be identical with those of the present occupant. If it is vacant, the properties of the new element can be predicted with the usual degree of certainty from what is known of the other members of the same group.

X-ray Spectra.—Such a discussion as that upon which we are engaged would be incomplete without an account of certain recent studies of X-ray spectra which have thrown unexpected light upon both the arrangement of atoms in crystals and upon the relationships of the periodic system.

The stimulus to these investigations began with an observation made by Laue in 1913. He found that when a narrow pencil of X-rays is allowed to pass through a crystal, and then to strike perpendicularly upon a photographic plate, a dark central shadow is formed by the beam at the point of contact, but this is surrounded by symmetrically arranged dots which were recognized by Laue as due to diffraction. Now ever since the discovery of the X-rays a dogma had prevailed to the effect that they could be neither reflected, refracted or diffracted. The fact is, however, that no diffraction grating can be efficient unless the spacing of its lines is of the same order of magni-

ıde as the wave lengths of the vibrations concerned. Now
:-rays have wave lengths approximately only one ten-thousandth
hat of sodium light. This would require a grating the dis-
ance between whose lines was as small as that between the mole-
ulés in a crystal. It therefore occurred to W. H. Bragg of Leeds
nd W. L. Bragg of Cambridge to use these spaces between mole-
ules for this very purpose, and they found by experiment that
hen a beam of X-rays strikes at an angle upon the face of a
rystal, a kind of reflection occurs in which several surface
ıyers take part. It is somewhat analogous to the way in which
n opalescent substance reflects light. What happens is best
tated in their own words:

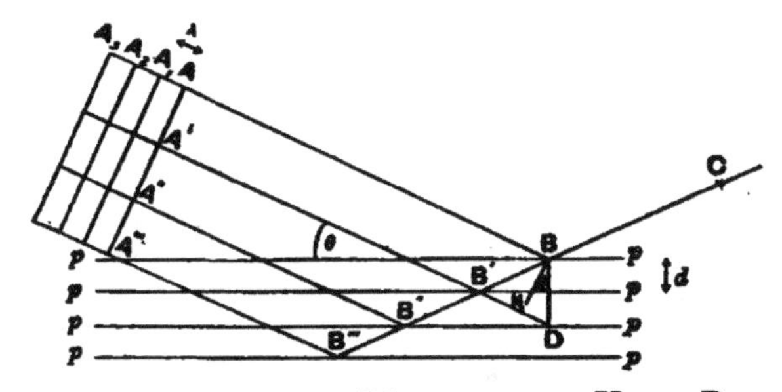

)IAGRAM ILLUSTRATING THE MECHANISM OF X-RAY REFRACTION
eproduced from W. H. and W. L. Bragg's "X-rays and Crystal Structure"
by the kind permission of the authors and of G. Bell and Sons.

"Let the crystal structure be represented by the series of planes
,*p*,*p*; *d* being their common distance apart or 'spacing.' A, A_1, A_2, A_3,
. . are a train of advancing waves of wave length λ. Consider
ıose waves which, after reflection, join in moving along *BC*, and
ɔmpare the distances which they must travel from some line such as
A''' before thay reach the point *C*. The routes by which they travel
:e *ABC*, *A'B'C*, *A''B''C*, and so on. Draw *BN* perpendicular to
'*B'*. Produce *A'B'* to *D*, where *D* is the image of *B* in the plane
ırough *B'*. Since $B'B = B'D$, and $A'N = AB$, the difference between
'*B'C* and *ABC* is equal to *ND*, that is, to $2d \sin \theta$. Similarly, *A''B''C*
greater than *A'B'C* by the same distance and so on.

"If *DN* is equal to the length of the wave, or is any whole multiple
: that length, all the wave trains reflected by the planes *p*,*p*,*p*, are in
ıe same phase and their amplitudes are added together. If *DN*
iffers but slightly from the wave length, say by a thousandth part,

the many thousand reflections bear all sorts of phase relations to each other, and the resultant amplitude is practically zero. We see, therefore, that when a monochromatic wave train is allowed to strike the face of the crystal, it is only when the glancing angle has certain values that reflection takes place. These values are given by

$$\lambda = 2d \sin \theta_1$$
$$2\lambda = 2d \sin \theta_2$$
$$3\lambda = 2d \sin \theta_3, \text{ etc.}$$

"The reflection at the angle θ_1 is called the reflection of the first order, that at the angle θ_2 reflection of the second order, and so on.

"If the crystal is slowly turned round in such a way that the glancing angle steadily increases, in general there is no reflected beam. But as the angle assumes the values θ_1, θ_2, θ_3, there is a reflection of the rays. Passing now to another face of the crystal which has a different spacing, d', the monochromatic rays will only be reflected when

$$\lambda = 2d' \sin \theta_1$$
$$2\lambda = 2d' \sin \theta_2, \text{ etc.}$$

"If, therefore, we measure the angles θ_1, θ_2, θ_3, at which reflection occurs, it gives us a relation between λ, the wave length, and d, the constant of the grating. By employing the same crystal face the wave lengths of different monochromatic vibrations can be compared. By using the same wave length, the distance d can be compared for different crystals and different faces of the same crystal."

The Atomic Structure of Crystals.—To test this reasoning the authors constructed an X-ray spectrometer. In its arrangement this resembles a reflecting goniometer, but the rays after being reflected by a crystal pass into an ionizing chamber where their intensity can be measured by the conductivity which they impart to the gas it contains. It is clear that if a given radiation consisted of vibrations of all wave lengths, there would be some ionization in the chamber for every position of the crystal. This would correspond to a continuous spectrum. If, on the other hand, the radiation consisted of a few monochromatic rays, then at certain angles a marked increase of ionization would suddenly result. When plotted these would correspond to a line spectrum. As a matter of fact there are both effects. There is a weak and characterless continuous spectrum, and, at certain angles, there are groups of two or three lines which are usually close together and which are characteristic of the source of the

It will be remembered that in the ordinary X-ray bulb athode rays generate the X-rays by striking a tilted plate of l, usually platinum or tungsten, which is called the anti-de. Now it has been known for some time that anti-des of different material produce different kinds of rays, and la, using a great variety of anticathodes, has found that the ing rays differ widely in their coefficients of absorption. e experiments of W. H. and W. L. Bragg it is clearly demon-d that they have different spectra and from this can be n some remarkable conclusions concerning the atoms which them. Before discussing this latter point, however, we come back to the fundamental equation:

$$\lambda = 2d \sin \theta.$$

obviously contains two unknown quantities λ and d, but we some data for determining at least the order of magnitude e latter. If we know the mass of an atom of hydrogen 259), the molecular weight of the substance, and its den-it is an easy matter to calculate the average space occu-by each molecule. To get an accurate measure of the wave h, however, it is necessary to have a more accurate value, in a crystal it is to be expected that the distances be-particles will vary with the direction. It has proved ole to accomplish this by the study of crystals of simple ical composition and highly symmetrical form. Rock salt crystallizes in the cubic system is admirably adapted for urpose. By a somewhat elaborate comparison of geometrical ls too extensive to be repeated here, the Braggs prove that cubic system the structure of any crystal must conform to f three space lattices, and they show that these can be dis-ished by experiment, for if X-ray reflection be made on incipal crystal faces {100}, {110}, and {111} the sines of the s of first order reflection will have different ratios in the cases. These are:

For the cube lattice, $1 : \sqrt{2} : \sqrt{3}$

For the cube-centered lattice, $1 : \frac{1}{\sqrt{2}} : 3$

For the face-centered cube lattice, $1 : 2 : \frac{\sqrt{3}}{2}$

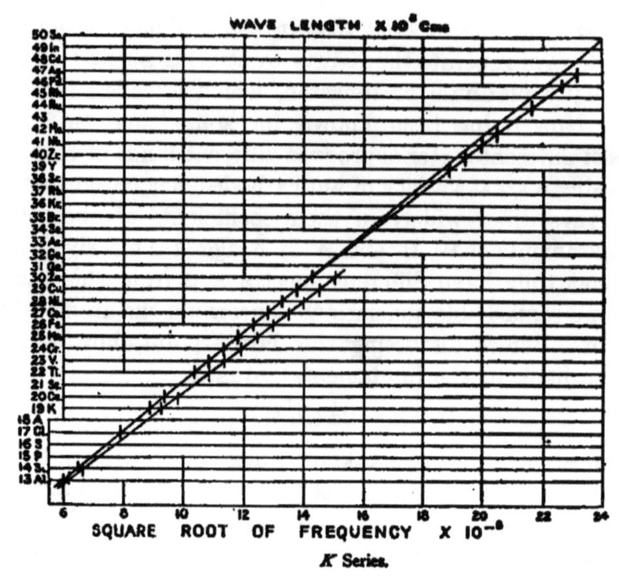

K Series.

RELATIONSHIP OF X-RAY SPECTRA AND ATOMIC NUMBERS

Reproduced from W. H. and W. L. Bragg's "X-rays and Crystal Structure" by the kind permission of the authors and of G. Bell and Sons.

Now it happens that KCl, NaCl, FeS and CaF_2 all crystallize in the cubic system and their space lattices are such that in each case one-half of a molecule is associated with what may be designated as the "unit cube." In any of these cases, therefore, the spacing between the particles can be found from the equation:

$$(\tfrac{1}{2} M)m = \rho(d^3_{(100)})$$

in which M is the molecular weight, m the mass of an atom of hydrogen, ρ the density of the crystal, and $d_{(100)}$ the space between the layers of particles parallel to the crystal face $\{100\}$. When the values of d obtained in this way from all four of the substances mentioned are substituted in the fundamental equation we obtain the same value for λ, which is a valuable check upon the correctness of the assumptions involved.

It has been pointed out above that with X-rays of a given wave length it is possible to investigate the structures of various crystals by determining the depth between layers parallel to known

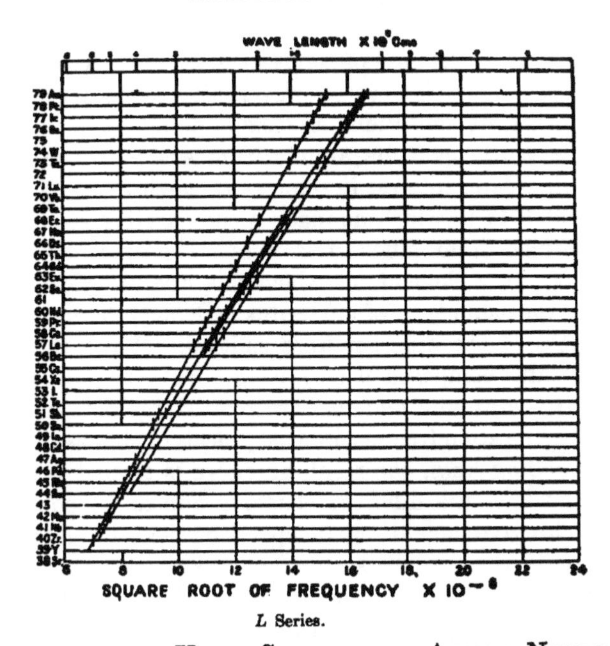

L Series.

RELATIONSHIP OF X-RAY SPECTRA AND ATOMIC NUMBERS

Reproduced from W. H. and W. L. Bragg's "X-rays and Crystal Structure" by the kind permission of the authors and of G. Bell and Sons.

crystal faces. Several investigators are now carrying on this work, and have obtained interesting results. It is found, for example, that the reflection of the rays is a function exercised by the atoms of each element separately, regardless of the compounds in which they are combined, so that the arrangement of the atoms of a given element in a crystal can be studied independently of the other atoms present, just as a geometrical figure (to use Bragg's illustration) may be used to connect the recurring points of a wall-paper design, without regard to the other details of the pattern. This has sometimes been expressed in the form that in a crystal the molecule has no separate existence, since, in rock salt, for example, each atom of sodium stands in the same space relation to several atoms of chlorine. A still more important conclusion must be drawn from these studies. In a crystal, the atoms must be rigidly fixed, or at least they cannot have the latitude of vibration which has hitherto been assumed all states of aggregation. This is particularly interesting, as

T. W. Richards from his work on compressibility, and Pope and Barlow, from considerations of structure, have each come to the conclusion that in the case of solids at least, the atoms must be close-packed. There is as yet, however, little agreement between these investigators concerning matters of detail.

Mosely's Work upon the Atomic Numbers.—To the chemist there is still greater interest in some work carried out by Mosely in 1913 and 1914 in which he made a comparative study of the X-rays derived from different sources. He constructed anti-cathodes of every element for which this was practicable, and, employing the same type of apparatus as Bragg, determined the wave lengths of the different lines emitted. When the results are graphically compared a surprising regularity becomes apparent. If the atomic numbers are plotted against the square roots of the vibration frequencies (reciprocals of the wave length) of corresponding lines, the resulting curve is almost a straight line, upon which the different elements appear at equidistant points. This gives an entirely independent check upon the order of elements in the periodic system, and possesses certain advantages over every other periodic function hitherto studied. The results are reassuring. Three conclusions stand out prominently, and are worth emphasizing: first, the order of the existing elements is the same as that already adopted on the basis of chemical analogy, even where this contradicts the strict order of the atomic weights, as in the case of argon and potassium; second, the elements of the rare earth group all find separate places upon the curve, and are therefore entitled to similar recognition in the table, and cannot all be grouped in one place as has been done by some theorists; third, the fact that the elements in this arrangement are equidistantly spaced shows more clearly than has hitherto been possible, exactly the number of new elements whose discovery may be expected and their character. As a matter of fact there are now but three vacant spaces, so that the discovery of many more kinds of elements need not be looked for. This of course places no limit upon the possible number of isotopes, for these all have the same atomic number.

The most important of these conclusions is the first mentioned, which shows us that while they run closely parallel, the atomic number of an element is a more fundamental index of its quality an the atomic weight. This value must depend upon something sely allied to mass but not identical with it, and we must now tate the periodic law in the terms: The properties of the elents are periodic functions of their atomic numbers.

Literature

'he best books on radioactivity and allied subjects are:

. J. Thomson: *Conduction of Electricity through Gases*, 2d edition, Camlge, 1906 (extremely mathematical).

utherford: *Radioactivity* 2d edition, Cambridge, 1905. *Radioactive nsformations* 2d edition, New York, 1911.

oddy: *The Interpretation of Radium*, 3d edition, London 1909 (written opular language).

he Chemistry of the Radioelements, 2d edition, London, 1914.

. H. and W. L. Bragg: *X-rays* and *Crystal Structure*, 2d edition, lon, 1916.

INDEX

gures in **heavy type** indicate a biographical note or a full discussion.
Asterisks refer to illustrations

A

C

F

H

M

N

O

S

T

U

X

Z

CPSIA information can be obtained
at www.ICGtesting.com
Printed in the USA
BVHW030812031118
532088BV00009B/69/P